BASIC

GEOMETRY

Answer Book

<u>Pages 16-17</u>

1. No. (a) He headed always for the summit.
       (b) He looked toward the zenith.
2. The "light cream" is heavier. Thin; thick.
3. The sugar dissolves. The butter melts. Cold substances can melt in your mouth. The oil dissolves in the gasoline.
4. "Mad," often used to mean "angry," really means "mentally deranged."
5. An "ambiguous" statement can be taken to mean two different things.
6. It can mean "He deserves the best of everything" and also "His just desert is less than nothing."
7. They may never have actively helped him.
8. Yes, usually; but not at the South Pole.
9. South, also.   10. Vertical
11. No.   12. No.
13. Yes; consider the three edges of an ordinary box.
14. Yes.
15. Colloquially, yes. Strictly, you could "lay it down flat" but tilted.
16. No. It might be flat and also vertical. A level surface is perpendicular to a plumb line.

<u>Pages 19-20</u>

1. a; d, though some of the solid lead will melt eventually.

2. c; b, though some of the solid iron will melt eventually.

3. d; a, though some of the solid iron will melt eventually.

4. b;c, though some of the solid lead will melt eventually.

5. Solid ludge sinks in molten runk.

6. True in number system to base 5.

7. That the government chose to test the twelve best of all brands of weather-strip manufactured, and that each of the other eleven was rated less than 93% efficient.

Pages 24-25

1. Use Assumption 1.

2. $\angle ABD = \angle ACD$. Use Theorem A twice.

3. $\angle BAD = \angle CAD$. Use Ex. 2 and Assumption 1.

4. $\angle BAC = \angle BDC$. Draw AD and use Theorem A twice.

5. (d) Given, a chord and the radius perpendicular to it. (e) Given, a random triangle.

6. (a) If two chords of a circle are equal, then they are equally distant from the center of the circle. (b) If a quadrilateral is a parallelogram, then the opposite angles of the quadrilateral are equal. (c) If a baby is hungry, (then) it cries.

Pages 28-29

1. If all three angles of a triangle are equal, the three sides are equal also. True.

2. If a quadrilateral is a parallelogram, all four sides are equal. False.

3. If the angles of two triangles are respectively equal, the two triangles are equal. False.

4. If the diagonals of two rectangles are equal, the rectangles are equal. False.

5. If the grass is wet, dew has fallen. False.

8. If the tree has no sap in it, the tree is dead.
True.

11. Every point in line ABC is a point in line AB.
False.

Pages 30-31

1. "All roads going west lead to Indiana" does not necessarily imply that all roads to Indiana go west.

3. The dietary needs of David with typhoid fever and of David in normal health are not the same.

6. Quite possibly there is something about the way Arctic explorers live, and not the tea, that causes the deep lines.

7. If both pieces of steel suffered change, and proportionately, no change in measure would be observed.

8. The word "average" implies a 50-50 split, or almost.

Pages 41-42

1. $AB = 1.0$, $BC = 0.5$, $CD = 1.25$, $AC = 1.5$, $BD = 1.75$, $AD = 2.75$.

    (a) $1.0 + 0.5 = 1.5$
    (b) $1.0 + 0.5 + 1.25 = 2.75$
    (c) $2.75 = 1.0 + 1.75 = 1.5 + 1.25$

2. (a) $5/2$; (b) Yes
3. (a) 180; (b) 100
4. (a) 3/4 foot; (b) 6/7
5. (a) 9 inches; (b) 6/7
6. (a) 23 centimeters; (b) 137/160

Page 48

1. $60 = 30 + 10 + 20$
2. $60 = 40 - 10 + 30$
3. $80 = 40 + 30 + 70 - 30 - 10 - 20$

Pages 51-52

1. The half-line OM' has the number 270, so that

angle L'OM' has the measure 270 – 180, or 90°, and so that angle M'OL has the measure 360 – 270, or 90°.

    2. 90°; 60°; 60°; 300°; 360°

    3. 420°; 1350°; 228°

    4. 8:31

    5. 1:48 A.M.

    6. 243°

    7. 225 (or 225 ± n · 360)

    8. r + 180 or r - 180

    9. 132, 180, 312

  10. a + r, a + 180, a + r + 180

  11. 54.2°, 125.8°, 54.2°, 360°

  12. 180° - s°, s°, 180° - s°

  13. Angles AVD and BVC

  14. 90°

  15. $\angle$AVC + $\angle$CVB = 180° $\therefore$ 1/2 $\angle$AVC + 1/2 $\angle$CVB = = 90°.

  16. 360° – s°

<u>Page 53</u>

    1. 33 1/3 grades, 533 1/3 mils

    2. 150 grades, 2400 mils

    3. 22 1/2°, 400 mils

    4. 54°, 960 mils

    5. 5.625°, 6.25 grades

<u>Page 56</u>

    2. EA = 1· 5/8 in. (approx.); $\angle$DEA = 100° (approx.); $\angle$EAB = 99° (approx.)

    3. CA = 9.0 cm.; $\angle$BCA = 52°; $\angle$CAB = 59°. Sum = 181° (approx.)

    4. Approximately 14.6 miles

    5. 8 miles on second course; 29 1/4 miles in all.

<u>Page 60</u>

    1. 3/2

2. $C'A' = 3 \cdot CA$; $\angle B'C'A' = \angle BCA$

5. The third side of the second triangle is <u>not twice</u> the third side of the first triangle; the second and third angles of the second triangle are <u>not equal</u> to the second and third angles respectively of the first triangle.

6. Yes.

8. We may use either pair of diagonals.

9. Prove that the two quadrilaterals have corresponding angles equal and corresponding sides proportional.

16. Apply Principle 5 to the two right triangles.

17. See Fig. 21 on page 82 and text on upper half of page 83.

18. See first paragraph on page 82.

19. Choose a point C on the ground as shown in Fig. 30. On AC produced lay off CA' equal to AC; on BC produced lay off CB' equal to BC. Measure B'A'.

20. 1.5

21. 2/3

22. 0.2

23. $\dfrac{n - m}{m}$

24. 1/2

25. 1/2

26. Each fraction equals 2/1

27. $AB'/AB = BB'/AB + AB/AB = 1/2 + 1 = 3/2$

28. 5/3

29. 6/5

30. $\dfrac{n + m}{m}$

31. Multiply each side of the equation $a/b = c/d$ by bd; then divide each side by ac.

32. Multiply by bd, as in Ex. 31; then divide each side by cd.

34. Write b/a = d/c and add 1 to each side.

36. Subtract 1 from each side of the equation.

37. $\dfrac{a + c + e + \cdots}{b + d + f + \cdots} = \dfrac{k(b + d + f + \cdots)}{b + d + f + \cdots} = k = \dfrac{a}{b}$

38. Apply the theorem in Ex. 37.

Pages 68-69

1. AB = BA = 1.5
2. -- 1.5
3. 1.5; - 1.5
4. (5.1 - 2.7) + (4.2 - 5.1) = 4.2 - 2.7
5. (4.2 - 5.1) + (2.7 - 4.2) = 2.7 - 5.1
6. (2.7 - 5.1) + (4.2 - 2.7) = 4.2 - 5.1
7. Points B and C and all the points between B and C.

8. Points A and B and all the points between A and B.

9. Angle AOB = angle BOA = 69°
10. 55°; - 55°; 124°; - 124°
11. (161 - 106) + (37 - 161) = (37 — 106).
12. 127, or 307, or ...
13. About 2 miles SE of Toggenburg.
15. 1/3 and 3/4

Pages 73-75

1. By Principle 6
2. 3/2 as long; $\angle$C′ = $\angle$C
3. KL = 1.3 K′L′; LM = 1.3 L′M′
5. 2a
6. 25 feet
7. 17/24 x 5 feet, or 3 ft. 6 1/2 in.
8. Extend CA its own length and erect a perpendicular; or, if C is not a right angle, copy $\angle$C.
9. Approximately 59° 2′ or 59.04°
10. Triangles ABE and ACF are similar, because two angles of one are equal to two angles of the other.

Therefore h/c = k/b, and bh = ck.

11. If $\angle$C = 90°, D and E coincide with C. If $\angle$C > 90°, triangles ABE and ACF are still similar.

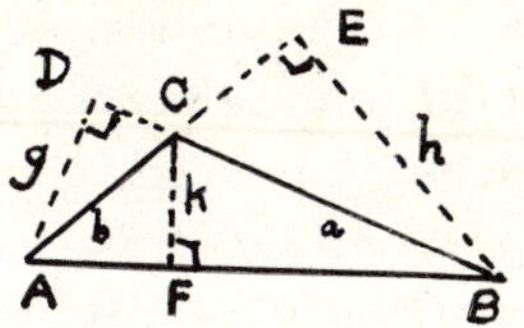

12. Use similar triangles ABD and BCF to give
g/c = k/a, whence ag = ck.

13. No

14. No

Pages 77-78

3. $\angle$ABC = $\angle$ABD + $\angle$CBD = $\angle$ADB + $\angle$CDB = $\angle$ADC.

4. Triangles ABC and ADC are similar by preceding exercise and Principle 5. The factor of proportional-
ity is 1. Therefore $\angle$BCA = $\angle$DCA. Hereafter, when two
triangles are equal (see p. 59) we shall write simply
$\triangle$ABC = $\triangle$ADC. In <u>Basic Geometry</u> the symbol $\triangle$ for
triangle does not appear until page 200.

5. $\angle$KBC = $\angle$CBD − $\angle$KBD = $\angle$CDB − $\angle$KDB = $\angle$KDC

6. $\angle$BAC = 180° − p = 180° − q = $\angle$BCA. Therefore
AB = BC by Principle 7.

7. $\angle$BAC = $\angle$BDC by Ex. 3; $\angle$ABC = $\angle$DBC by Ex. 4;
$\angle$BAD = $\angle$BDA by Principle 7. Therefore $\triangle$ABE = $\triangle$DBE
by Principle 6, and $\angle$AEB = $\angle$DEB. ∴ $\angle$AEB = 90° =
$\angle$DEB by Ex. 14 on p. 52. For definition of perpendicular,
see p. 50.   8. Yes.

Pages 80-82

1. The angles in the triangles will remain unchanged.
Angles ABC, DCB, EFG, HGF and the angles of the quadrilaterals can vary.

2. The nails insure that the three sides of the
triangle remain unchanged. Then by Principle 8 the angles
of the triangle cannot change.

3. The rigidity of a triangle no one of whose sides
can change in length (Principle 8).

4. Five distinct cross-braces can be made. Only two of these (any two) are needed for rigidity. These two can be chosen in ten different ways. For, the first diagonal brace can be chosen in five ways, the second in four ways; but of these twenty, half the patterns duplicate the other half. (Or, the number of distinct combinations of five things taken two at a time is $(5 \cdot 4)/2$.)

5. $\angle A = \angle B$ by Principle 7.

$\angle ACM = \angle BCM$ by Principle 5.

$\angle AMC = \angle BMC$ by Principle 5; and each of these angles equals 90° by Ex. 14 on page 52.

6. The theorem and the converse theorem have been proved in Ex. 16 on page 63 and in Ex. 5 on page 81.

7. Each angle of the larger triangle equals 90°.

8. No.

Pages 84-87

1. All four triangles are similar to $\triangle ABC$, with k equal to 1; three of them by Principle 5, and one by Principle 8.

2. $\angle A + \angle B + \angle ACB = 180^\circ$ by Principle 9.

$\angle BCD + \angle ACB = 180^\circ$.

$\therefore \angle BCD = \angle A + \angle B$

3. By drawing a diagonal the quadrilateral can be divided into two triangles whose six angles make up the four angles of the quadrilateral. The sum of the six angles is 2 times 180°, or 360°.

4. Similarly, by drawing two diagonals from the same vertex, a pentagon can be divided into three triangles whose nine angles make up the five angles of the pentagon. The sum of the nine angles is 3 times 180°.

5. Again, a hexagon can be divided into four triangles, the sum of whose angles is 4 times 180°.

6. Any convex polygon of n sides can be divided into $(n - 2)$ triangles the sum of whose angles is $(n - 2)$ times 180° and equal also to the sum of the angles of the

polygon. Or, the polygon can be divided into n triangles as shown in Fig. 21 on page 59, and we must subtract 2 times $180°$ from n times $180°$ to get rid of the n angles with common vertex at the interior point of the polygon.

7. $(3 \times 180°) \div 5 = 108°$

8. $(4 \times 180°) \div 6 = 120°$

9. $(6 \times 180°) \div 8 = 135°$

10. $\dfrac{n-2}{n} \cdot 180°$

11. Set $\dfrac{n-2}{n} \cdot 180°$ equal to $144°$; $n = 10$

12. By Principle 6.

13. The other pair of adjacent angles are equal, by Principle 9. Then the two triangles are similar, with k equal to 1, by Principle 6.

14. Using Fig. 25, right triangle ACD is similar to right triangle ABC, and right triangle CBD is similar to right triangle ABC, by Ex. 12 on page 85. Hence right triangles ACD and CBD are similar, and $m/h = h/n$, or $h^2 = mn$.

15. By Principle 6, triangles ACD and ABC are similar. Therefore $m/b = b/c$, or $b^2 = cm$. Also, triangles CBD and ABC are similar, and $n/a = a/c$, or $a^2 = cn$.

16. $h^2 = 5 \times 8$ and $h = 2\sqrt{10}$; $a^2 = 8 \times 13$ and $a = 2\sqrt{26}$; $b^2 = 5 \times 13$ and $b = \sqrt{65}$.

17. The similarity of the three right triangles tells us first that $m = \dfrac{5}{12}h$, $n = \dfrac{12}{5}h$, and $m+n = \left(\dfrac{5}{12} + \dfrac{12}{5}\right)h = \dfrac{169}{60}h$. Then, using $b^2 = m(m+n)$ as in Fig. 25, we have $25 = \dfrac{5}{12}h \times \dfrac{169}{60}h$; whence $h^2 = \dfrac{25 \times 144}{169}$, and $h = \dfrac{60}{13}$. Then $m = (5/12)h = 25/13$ and $n = (12/5)h = 144/13$.

18. Using Fig. 25, and taking $m = 1 \cdot k$ and $n = 3 \cdot k$,

we have $\dfrac{a}{b} = \dfrac{h}{m} = \dfrac{\sqrt{3}\cdot k}{1\cdot k} = \dfrac{\sqrt{3}}{1}$.

19. Using Fig. 25, and taking $m = 1\cdot r$ and $n = k\cdot r$, we have $\dfrac{a}{b} = \dfrac{h}{m} = \dfrac{\sqrt{k}\cdot r}{1\cdot r} = \dfrac{\sqrt{k}}{1}$.

20. Using Fig. 25, and taking $m = p\cdot k$ and $n = q\cdot k$, we have $\dfrac{a}{b} = \dfrac{h}{m} = \dfrac{\sqrt{pq}\cdot k}{p\cdot k} = \sqrt{\dfrac{q}{p}}$. Or, more briefly, replace the ratio 1 to k in Ex. 19 by the ratio p to q, so that $\dfrac{1}{k} = \dfrac{p}{q}$ and $\dfrac{\sqrt{k}}{1} = \sqrt{\dfrac{q}{p}}$.

21. Using Fig. 25, we have $\dfrac{m}{h} = \dfrac{b}{a}$ and $\dfrac{n}{h} = \dfrac{a}{b}$, whence $\dfrac{m}{n} = \dfrac{b^2}{a^2}$. Also, from $\dfrac{h}{a} = \dfrac{b}{m+n}$ we get $h = \dfrac{ab}{\dfrac{b}{a}\cdot h + \dfrac{a}{b}\cdot h}$

and $h^2 = \dfrac{ab}{(b^2+a^2)/ab} = \dfrac{a^2 b^2}{a^2+b^2}$, so that $h = \dfrac{ab}{\sqrt{a^2+b^2}}$.

22. In a polygon of n sides there is an interior angle and an exterior angle at each vertex, and the sum of the two is $180°$. The sum of the exterior angles is

$n\cdot 180° -$ (the sum of the interior angles), or
$n\cdot 180° - (n-2)\cdot 180°$;

that is, $2\cdot 180°$.

23. $360° - (42 + 61 + 40 + 53 + 87 - 74 + 69) =$
$= 360 - (143 + 53 + 13 + 69) =$
$= 360 - (278) = 82°$.

<u>Pages 95-98</u>

1. (a) 5; (b) 35; (c) 52: (d) $\sqrt{2}$, or 1.4; (e) $s\sqrt{2}$; (f) $\sqrt{3}$; (g) 2; (h) 2s; (i) $\sqrt{5}$; (j) $\sqrt{6}$; (k) $\sqrt{7}$; (l) $\sqrt{n+1}$; (m) $p^2 + q^2$; (n) 16.6, or 17; (o) 11.3, or 11; (p) 486.

2. (a) 15; (b) $\sqrt{3}$, or 1.7; (c) 1; (d) 1; (e) $s\sqrt{3}$; (f) s;

(g) $\frac{s}{2}\sqrt{3}$; (h) s; (i) 12.4, or 12; (j) 8.7; (k) 171.

3. 20.2, or 20, miles.

4. In Fig. 38, prolong BD to C so that BD = DC. Then $\triangle$ADC = $\triangle$ADB and $\angle$C = $\angle$BAC = 60°. Then BC = BA by Principle 7 and BD = 1/2 BA. Alternatively, construct $\angle$KLM = 60°. Then, by Principle 9, $\angle$KML = 60° and $\angle$MLH = 30° = $\angle$H. It follows from Principle 7 that KL = KM and that KM = ML = MH. Hence KL = 1/2 KH.

5. In Fig. 38, assume BD = 1/2 BA and prolong BD to C so that BD = DC. This makes BC equal to BA, and $\angle$B = $\angle$C = $\angle$BAC by Principle 7. Then each of these three angles equals 60°, by Principle 9. Alternatively, from the mid-point M of HK draw MN perpendicular to KL and meeting KL at N. Then KN = 1/2 KL by Principle 6, and $\triangle$KMN = $\triangle$LMN by Principle 5. We are given that KL = 1/2 KH = KM, and KM = ML by Principle 5. Hence KL = KM = ML and $\angle$K = 60°.

6. The shortest side and the hypotenuse are in the ratio 1 : 2. Suppose their lengths to be 1k and 2k, respectively. Then by Principle 12 the third side will be $\sqrt{3}$k in length.

7. Construct a right triangle with sides including the right angle having lengths 1k and $\sqrt{3}$k. Then the hypotenuse must be 2k. By Principle 8 this triangle is similar to the given triangle. So the given triangle must be a right triangle. Exercise 5 shows that this must be a 30°-60° right triangle.

8. Since $5^2 + 12^2 = 13^2$, we know from Principle 12 (converse) that $\angle$ABC = 90°.

9. Since $6^2 + 6^2 = (6\sqrt{2})^2$ we know from Principle 12 that $\angle$E = 90°. Principles 7 and 9 tell us that $\angle$D = $\angle$F = 45°.

10. By Ex. 7, $\triangle$GHK is a 30°-60° right triangle.

11. 5 inches; 12.4 inches; 12.6 inches.

12. 13 inches.

13. Use Principle 12 (converse), noting that
$$(p^2 - q^2)^2 + (2pq)^2 = (p^2 + q^2)^2.$$

14. $p = 2$, $q = 1$.

15. $p = 3$, $q = 2$.

16. 3, 4, 5; 5, 12, 13; 7, 24, 25; 8, 15, 17; 9, 40, 41; 11, 60, 61; 12, 35, 37; 30, 21, 29; 28, 45, 53; 33, 56, 65.

17. Enlarge this triangle so that the sides are 5 times as long; then so that the sides are 12 times as long. Place the two enlarged triangles together, as on p. 91. Then compare this composite triangle with the enlargement of the given triangle in which each side of the new triangle is x times as long as it was in the given triangle. Evidently $x^2 = 169$, and the hypotenuse $x = 13$.

18. In Fig. 36 think of BD as "the line." Then $AD < AB$, because AB is perpendicular to BD, by Corollary 12d. Similarly $DC < BC$, because DC is perpendicular to BD. Therefore $AD + DC < AB + BC$, or $AC < AB + BC$.

19. From Ex. 15 on p. 86 we have $b^2 = cm$ and $a^2 = cn$. Therefore $b^2 + a^2 = c(m + n) = c^2$. This proof requires first of all that the perpendicular CD in Fig. 25 on p. 86 shall be the only perpendicular from C to AB. This follows from Principle 11, which we proved by means of Principles 5, 10, and 9. Principle 10 depended on Principles 8 and 5; and Principle 9 depended on Principles 8 and 5. Hence Principle 11 depended on Principles 8 and 5. In order to prove Principle 8 we needed Principles 7 and 5; to prove Principle 7 we needed Principles 6 and 5; and to prove Principle 6 we needed Principle 5. So this method of proving the Pythogorean Theorem depends ultimately upon Principle 5.

The method of proof that we employed on pages 92-3 required Principles 5, 6, 9, and 8; hence 5, 6, and 8; hence

5, 6, and 7; hence 5 and 6; and ultimately only Principle 5.

20. From Corollary 12c, AB + BC $>$ AC. Subtracting BC from each side of this inequality gives AB $>$ AC − BC.

21. $\angle$DEP + $\angle$DPE = 90° = $\angle$DFP + $\angle$DPF.

But $\angle$DPE $<$ $\angle$DPF

Therefore $\angle$DEP $>$ $\angle$DFP.

22. In Fig. 39 the similar triangles with sides 1, 1, x and x, x, 2 yield the proportion 1/x = x/2, whence x = $\sqrt{2}$.

23. x^2, area of inner square, equals 1/2 of 2^2, area of outer square.

Pages 100 -103

1. 33/19 x 4 inches

2. 7/5 x 22 cm

3. 7 1/2, 7 1/2, 6

4. Take $\angle$AOB = 30°, OB = 3/16 in.; $\angle$AOC = 60°, OC = 3/8 in.; and so forth.

5. The angle at each vertex of a regular hexagon is 120°. The obtuse isosceles triangles in the accompanying figure have two angles equal to 30°.

6. Since the angle at each vertex of a regular pentagon is 108°, the isosceles triangles formed by prolonging the sides of the pentagon have angles 72°, 72°, 36° and each angle of the star is 36°.

7. 60°

8. The angle at each vertex of the n-gon is $((n-2)/n) \cdot (180°)$. Hence the isosceles triangle formed by prolonging the sides of the n-gon has angles 360°/n, 360°/n, and 180° − 4·(180°)/n. This last angle can be written as $((n-4)/n) \cdot (180°)$.

9. Use Principle 6 to prove that the foot of each perpendicular is also the mid point of a side of the triangle.

10. Follows at once from Ex. 9.

11. From Ex. 9 we know that the mid-point of the hypotenuse lies on the perpendicular bisector of each of the other two sides. Hence it lies on both perpendicular bisectors and is their point of intersection.

Alternative proof: The perpendicular bisector of BC meets the hypotenuse at I, such that $\triangle$IBM is similar to $\triangle$ABC (by Principle 6) and IB = 1/2 AB. Similarly, the perpendicular bisector of AC meets the hypotenuse at J, such that AJ = 1/2 AB. Therefore I and J coincide at the mid-point of the hypotenuse.

12. The diagonals of each face are $s\sqrt{2}$. The diagonals of the cube are $s\sqrt{3}$.

13, 14. Any pair of diagonals of the cube are the hypotenuses of equal right triangles. Thus AG is the hypotenuse of triangle ACG; EC is the hypotenuse of triangle CAE; HB is the hypotenuse of triangle BDH. In each of these triangles the sides including the right angle have lengths e and $e\sqrt{2}$, whereas the corresponding right triangles in square ABCD are isosceles right triangles. The diagonals AC and BD of the square ABCD intersect at right angles. Diagonals AG and CE of the cube cannot intersect at right angles because angles GAC and ECA, though equal, are less than $45°$.

15. In other words, find the angle between the diagonals AG and EC of the rectangle ACGE in which AC = $e\sqrt{2}$, CG = e, and AG = EC = $e\sqrt{3}$. Tan $\angle$GAC = $\sqrt{1/2}$. Therefore $\angle$GAC = $35.3°$ $(35°18')$ and the desired angle is approximately $70.5°$.

16. Of all points P in face ABCD the point A is farthest from C, hence also farthest from G. Therefore, of all possible angles GPC, angle GAC is the least.

17. Height of house = $18 + 6\sqrt{5}$. $(AB)^2 = (18 + 6\sqrt{5})^2 +$ + 1744 = 2248 + 216(2.236) = 2248 + 483⁻; and AB = 52.3⁻.

18. The cosine of $\angle APQ$ = 2/3, so that $\angle APQ$ =
= 48.2^o (48°11$'$).

19. $2\sqrt{7}$ = 5.3 inches

20. $1/2\sqrt{41}$ = 3.2 inches

21. 500 x 5/6 = 416 2/3

Pages 113-117

1. Through K draw line n$'$ making with TV an angle
equal to $\angle$CHT. Treat n$'$ as m$'$ was treated in the last
paragraph of p. 110.

2. The angles marked as equal in Fig. 5 are so
because of Theorem 15. Therefore triangles ABC and
ADC are equal, by Principle 6, and AB = CD and AD = BC.

3. Use Theorem 15 as in Ex. 2, and Principle 9.

4. In Fig. 6 AB = CD from Ex. 2. Use Principle 6.

5. Draw one diagonal and prove the triangles equal
by Principle 8. Considering the diagonal now as trans-
versal, use pairs of equal angles in the two triangles to
prove that opposite sides of the quadrilateral are parallel
(Theorem 14).

6. Draw one diagonal and prove the triangles equal
by Theorem 15 and Principle 5. This makes the other
sides of the quadrilateral equal. Then apply Ex. 5.

7. Since the sum of all four angles of the quadrilateral is 360°, by Ex. 3 on p. 84, the sum of two adjacent
angles is 180°. This makes a pair of opposite sides
parallel. See note, bottom of p. 109.

8. A pair of equal triangles (Principle 5) makes a
pair of corresponding angles in the two triangles equal and
implies (by Theorem 14) that two opposite sides of the
quadrilateral, known to be equal by Principle 5, are also
parallel.

9. Follows at once from Principle 10.

10. The diagonals bisect each other (by Ex. 4) and
are given perpendicular. Hence each is the perpendicular
bisector of the other. Then, by Principle 10, all four

sides are equal.

11. Since opposite sides of the parallelogram are equal, and opposite angles equal also, the two pairs of triangles are equal, by Principle 5. Consequently EH = FG and EF = HG and EFGH must be a parallelogram (by Ex. 5).

12. JK = 1.8; KL = 2.7.

13. HJ = 1.4 x KL

14. HJ = JK = KL

15. All the acute angles are equal and all the obtuse angles are equal. All distances between rails of the same track are equal, and equal for all four tracks. If both double tracks are equally spaced, there are more equal distances.

16. $AB = r \cdot HJ = s \cdot H'J'$, or $AB/HJ = r$ and $HJ/H'J' = s$. Therefore $AB/H'J' = rs$; and since $AB = H'J'$, it follows that $rs = 1$, and r and s are reciprocals.

17. Merely replace AH and BJ in Fig. 4, page 111, by AJ and BH.

18. Use Case 2 of Similarity.

19. By drawing A'A and extending through A, show that $\angle A' = \angle A$. Similarly for B'B and C'C. Therefore the triangles are similar, and $B'C' = (8/5) \cdot 4$ and $C'A' = (8/5) \cdot 6$.

20. First method: Use Theorem 15.

Second method: Draw a second parallel through B and apply Theorem 16.

21. Use Case 1 of Similarity and Theorem 14.

22. Use Principle 5 and Theorem 14.

23. Either prove the upper and lower triangles similar, or draw a parallel through the intersection of the diagonals and apply Theorem 16.

24. Use Ex. 22 on p. 115.

25. By means of equal angles prove that triangle ABR is isosceles. Use Ex. 20, page 115.

26. By means of equal angles prove that triangle ABQ

is isosceles.

27. Obvious from Fig. 16.

28. By means of Ex. 22 prove two of the lines equal
and parallel; these, therefore, are opposite sides of a
parallelogram, by Ex. 6.

29. Take the random line in the plane of the parallelogram. Use Ex. 24 on p. 116. The length of the perpendicular drawn to the random line from the point of intersection of the diagonals is equal to one-fourth of the
sum of the perpendiculars from the vertices.

Page 121

1. The suggestion " - or any other convenient
distances as the unit - " is meant to imply that printed
squared paper is not necessary for these few exercises
and that the pupil can draw his own network in each case.

2. Slopes are 5/3; 2/4; 3/2. 11/26.

4. $OA = \sqrt{34}$ $OH = 2\sqrt{5}$

    $OB = 2\sqrt{5}$ $OI = \sqrt{29}$

    $OC = \sqrt{13}$ $OJ = \sqrt{65}/2$ or $\sqrt{16.25}$

    $OD = \sqrt{797}/4$ $OK = \sqrt{74}/2$ or $\sqrt{18.5}$

    $OE = OF = OG = \sqrt{34}$

5. $AC = \sqrt{5}$ $JK = \sqrt{13}/2$

    $BD = \sqrt{109}/4$ $ID = \sqrt{1285}/4 = 8.96$

    $AG = 8\sqrt{2}$ $KD = \sqrt{2041}/4 = 11.3$

    $EF = \sqrt{136}$ $JD = \sqrt{1781}/4 = 10.5^{+}$

    $EK = \sqrt{130}/2$ $CJ = \sqrt{233}/2$

6. Slope is $-1/2$.

7. Slopes are 2; -1; $-5/9$.

8. Slope of GK is 1/3; slope of BD is 3/10. The
slope of HB is 0.

9. IA has steepest slope, 10. GE, though inclined
more steeply to the x-axis than IA, has no slope.

Page 124

1. $y/x = 4/5$.   2. $y/x = -3/2$.

3. ay = bx, or bx − ay = 0.

<u>Pages 126-130</u>

1. The seven figures have 2, 5, 6, 4, 1, 2, 0 axes of symmetry respectively. All but the second and fifth have symmetry with respect to a point. The fourth, fifth, sixth show close relation to a network.

2. The left leaf, or leaflet, usually exhibits left-handed asymmetry; the middle leaf, or leaflet, is symmetric; and the right leaf, or leaflet, usually shows right-handed asymmetry.

3. Use Theorem 15 and Principle 6.

4. $PB/PB' = AB/A'B' = BC/B'C' = QB/QB'$
   Therefore $(PB'/PB) - 1 = (QB'/QB) - 1$, and
   $BB'/PB = BB'/QB$.

Since $PB = QB$, P and Q coincide and the three lines are concurrent. If $AB = A'B'$ and $BC = B'C'$, the three lines are parallel.

5. Use Theorem 15 and Principle 6.

6. Using Fig. 30, assume that AA' and CC' meet at P and that AA' and BB' meet at Q.

$$PA/PA' = AC/A'C' = AB/A'B'$$
$$QA/QA' = AB/A'B'$$

Therefore $PA/PA' = QA/QA'$; $(PA/PA') - 1 =$ $= (QA/QA') - 1$; and $AA'/PA' = AA'/QA'$. Hence P and Q coincide, and the three lines are concurrent.

7. If the triangles are equal, AA', and BB', and CC' are parallel, by Ex. 6 on page 113.

8. Using Fig. 31, follow the proof of Ex. 6 except that A', B', C' are now replaced by A'', B'', C'' and that we now write $(PA/PA'') + 1 = (QA/QA'') + 1$.

9. As indicated in Fig. 32, draw two random lines through P to meet <u>1</u> and <u>m</u>. Complete the triangle as shown and draw another triangle similar to the first so that the two triangles have their sides respectively parallel. The line joining P and the corresponding vertex in

the second triangle is the desired line, by Ex. 6 above.

10. 15/4 miles an hour.

11. The perpendicular bisectors of opposite sides of a rectangle coincide. The perpendicular bisectors of adjacent sides meet in a point that is equidistant from all four vertices of the rectangle. Since the diagonals of a rectangle are equal (Principle 5) and bisect each other (p. 113, Ex. 4), their intersection also is equidistant from all four vertices of the rectangle and hence must lie on the perpendicular bisector of each side of the rectangle.

12. If all three planes are perpendicular to a fourth plane and no two of the three planes are parallel, they intersect two at a time in three parallel lines. If two of these three planes are parallel, the three planes intersect in two parallel lines. If all three of these planes are parallel, they have no point in common.

13. Call the two given parallel lines l and m. If the "other" line, m, and the plane containing l have a point in common, this point will lie not only in this plane but in the plane determined by l and m. That is, it will lie on the intersection of these two planes, namely l. But this would mean that a point of m lies also on l, which is impossible. So m and the plane containing l can have no point in common.

14. If the two planes have a point in common they also have a line in common. If any point of this line be joined to the ends of that segment of the given perpendicular line that is included between the two planes, the resulting triangle will contain two angles of $90°$, which is impossible.

15. If the two lines of intersection have a point in common, this point must be common to the two parallel planes, which is impossible.

16. Given lines l and m in Fig. 33, join the point of intersection of l and the first plane with the point of intersection of m and the third plane, forming an auxiliary

line shown in the figure. Apply Theorem 16 to 1 and the
auxiliary line, and again to the auxiliary line and m.

17. Use Ex. 15 on this page, Theorem 15, Principle
6, and Ex. 5 on page 127.

18. If "the plane of these lines" is not parallel to
the given plane it has a line in common with the given
plane. This line of intersection cannot be parallel to both
the given intersecting lines; it must have a point in common with one of them. This point, therefore, must be
common to the given plane and to a line that is parallel
to the given plane. This is impossible.

19. One of the given lines and the parallel through
any point of it to the other given line determine a plane
that is parallel to "the other given line," by Ex. 13 on
page 129. Since there is only one such parallel through
any point of the first given line, there is only one such
parallel plane through this "any point." Further, this
plane contains the parallel through each of the other points
of the first given line.

20. Through the given point there are two lines one
of which is parallel to one of the given skew lines, while
the other is parallel to the other of the given skew lines.
These two "parallels" determine a plane, and the only
plane, that is parallel to both the given skew lines. If the
given point lies on one of the given skew lines we have the
situation of Ex. 19 on this page.

21. If there is a common perpendicular to two given
skew lines, it will be perpendicular also to a random plane
that is parallel to the two skew lines. So, of all the perpendiculars to a given skew line we need consider only
those that are perpendicular also to this random plane.
These perpendiculars lie in the plane that contains the
given skew line and is perpendicular to the random plane.
Similarly, we need consider only those perpendiculars to
the other skew line that lie in the plane that contains this
other skew line and is perpendicular to the random plane.

The line of intersection of these two planes each of which
is perpendicular to the "random parallel plane" is the
common perpendicular to the two skew lines.

Page 135

    1. (a) $x^2 + y^2 = 4$ (c) $x^2 + y^2 = 9.61$
       (b) $x^2 + y^2 = 25$ (d) $x^2 + y^2 = 4/9$

    2. (a) 3   (c) 1.9   (e) $\sqrt{2}$
       (b) $2\sqrt{2}$ (d) $3/2$

    3. One, the point (0, 0)

    4. None

    5. Arc BC = arc DE = 25   Arc BC = arc CE = 65
       Arc BE = arc FA = 90   Arc BF = arc EA = 235
          [Arc ED = arc CB = −335]

    6. On the half-line numbered 80, or 260.

Page 137

    1. Use Principle 5.

    2. Apply Case 3 of Similarity, pages 78-80, to the
two triangles.

    3. Use Corollary 12b, page 93.

    5. Use the Pythagorean Theorem and Ex. 3, page
137.

    6. Use the Pythagorean Theorem.

    7. Each chord corresponds to a central angle of
60°.

    8. First we need to show that the line of centers OO'
of two tangent dimes with centers
O and O' passes through the point
of tangency T. If OTO' is not a
straight line, then the length of
the straight-line segment OO'
will be less than the length of
OTO', by Corollary 12c. But
since two dimes, unlike two geo-
metric circles, cannot intersect in two points, we see that
the length of the straight-line segment OO' must also be

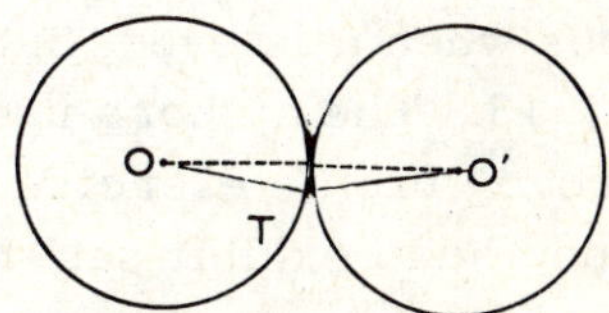

greater than 2r. This contradiction impels us to reject our original assumption that OTO' is not a straight-line segment. The rest is easy, following Ex. 7.

Pages 140-145

1. Use indirect method - suppose that perpendicular does not pass through the center; then Theorem 21, and unique perpendicular idea on page 54.

2. Use indirect method - suppose that the perpendicular does not pass through the point of tangency; then Theorem 21 and Principle 11, page 88.

3-4. Use Corollary 15a, page 111, and Corollary 12b, page 93.

5. Use Theorem 21, Corollary 15a, and Ex. 2 on page 141 to prove that the diameter through one point of tangency passes through the other point of tangency also.

6. Use Principle 8 to prove the equality of two triangles with common vertex at the center. Then use Principle 7 to prove that two adjacent angles of the parallelogram are not only "supplementary" (page 111) but also equal.

7. Use Ex. 3 and Theorem 19.

8. Use corollary 12b.

9. Use Ex. 8.

10. The sum of one pair of central angles is equal to the sum of the other pair. Use Ex. 8 and Principle 9.

11. As in Ex. 9.

12. As in Ex. 11, the sum of alternate sides of a circumscribed n-gon is equal to the sum of the remaining sides when n is even, but not when n is odd.

13. The authors use the word "show" instead of "prove" in this exercise to indicate that the pupil is expected to exhibit satisfactory diagrams only, but no proofs.

14. By Principle 10, page 88, and Principle 2, page 44.

15. If the two circles should have three distinct points, A, B, C in common, then OO' would be the perpendicular bisector of AB and of AC at the same time; and this is impossible, as there cannot be two lines from A perpendicular to OO'.

16. Use Ex. 1.

18. 3.99 inches

19. 0.87 inches

20. 4.8 + 4.4 + 4.0 = 13.2 inches

21. (a) $OO' > r + r'$ (c) $r - r' < OO' < r + r'$
    (b) $OO' = r + r'$ (d) $OO' = r - r'$
    (e) $OO' < r - r'$

22. OO'. If $r = r'$, a second axis of symmetry is the perpendicular bisector of OO' in cases (a), (b), and (c).

23. From a point P on the common tangent, a tangent to either circle is equal to PT, by Ex. 8 on page 141.

24. This is a special case of the preceding exercise.

25-26. If the common external tangents meet at T, then the angle between these tangents is bisected by TO (Ex. 8 on page 141). This same angle is bisected also by TO'. Therefore TO and TO' are (parts of) the same line, and T lies on OO'.

If the two circles in Ex. 25 have equal radii, their common external tangents do not meet.

part II

<u>Pages 147-152: Exercises.</u>

1. The sum of two opposite angles of the quadrilateral is equal to half the sum of two central angles that add up to 360°.

3. In Fig. 24 $\angle ABC = 90^\circ = \angle ABD$, so CBD is a straight line.

4. For the left-hand figure: $\angle C + \angle ABD = 180^\circ$ and $\angle ABD + \angle ABF = 180^\circ$.

Therefore $\angle C = \angle ABF$, $\angle ABF + \angle E = 180^\circ$.

Therefore $\angle C + \angle E = 180^\circ$ and chords CD and EF are parallel (by page 110, lines 1-3).

For the right-hand figure:

$\angle C + \angle ABD = 180^\circ$ and $\angle AEF + \angle ABD = 180^\circ$.

Therefore $\angle C = \angle AEF$ and chords CD and EF are parallel.

5. $\angle APC = \angle B + \angle C = \frac{1}{2} \angle AOC + \frac{1}{2} \angle BOD = \frac{1}{2}(\angle AOC + \angle BOD)$, where O is the center of the circle.

6. $\angle APC = \angle ABC - \angle BCD = \frac{1}{2} \angle AOC - \frac{1}{2} \angle BOD = \frac{1}{2}(\angle AOC - \angle BOD)$

7. Draw the bisector OM of the isosceles triangle TOB (Fig. 28) and prove that two angles at M are right angles. It follows that angle MOT and the angle between the tangent and the chord are both equal to $90^\circ - \angle MTO$.

9. Use the fact that the four angles of quadrilateral PSOT add up to 360° and that two of these angles are right angles.

10. From Ex. 9 $\angle SPT = 180^\circ -$ the lesser angle SOT =

$\frac{1}{2}(360^\circ -$ twice the lesser angle SOT$) =$

$\frac{1}{2}($greater angle SOT + lesser angle SOT $- 2 \cdot$ lesser angle SOT$) =$

$\frac{1}{2}($greater angle SOT - lesser angle SOT$)$.

11. $63\frac{1}{2}^\circ$

12. $180^\circ - 120\frac{1}{2}^\circ = 59\frac{1}{2}^\circ$

13. 69°

14. Note that Exs. 14, 17, 23, 24, and 25 say "show" - not "prove" - and "can be regarded." All that is expected of the student in these

five exercises is an intuitive recognition of limiting cases.

In Fig. 26, as D approaches B, $\angle C$ approaches 0° and $\angle APC$ approaches $\angle B + 0^\circ$.

15. $27\frac{1}{2}^\circ$

16. 34°

17. See note on Ex. 14.

In Fig. 27, as D approaches B, $\angle BCD$ approaches 0° and $\angle APC$ approaches $\angle ABC - 0^\circ$.

18. In Fig. 28, when $\angle TOB = 90^\circ$, $\angle OTB = 45^\circ$ = the angle between tangent and chord. When $\angle TOB = 180^\circ$, TB is a diameter and is perpendicular to the tangent at T.

19. $12\frac{1}{2}^\circ$

20. 52°

21. 43°

22. 222.2° and 137.8°

23. See note on Ex. 14.

24. See note on Ex. 14.

As B moves along the circle toward T, P moves toward T along the tangent and $\angle PAT$ approaches 0°.

25. See note on Ex. 14.

Let A and B withdraw from T along the circle until A and B approach coincidence.

27. Use Corollary 22c and Ex. 15, page 86.

28. $2\sqrt{3}$

29. $\dfrac{13}{4}$

30. In Fig. 31, let DO = x. Then AD = 4 - x, DB = 4 + x, and $(PD)^2 = 9 = (4 - x)(4 + x) = 16 - x^2$. Therefore $x^2 = 7$ and $x = \sqrt{7}$.

Alternative solution: Let AD = x. Then DB = 8 - x and $(PD)^2 = 9 = x(8 - x) = 8x - x^2$ and $x^2 - 8x + 9 = 0$. Using the quadratic formula,

$$x = 4 \pm \sqrt{7}.$$

31. $PA = 2\sqrt{6}$ and $PB = 2\sqrt{10}$

32. $PA = 4$ and $PB = 4\sqrt{3}$

33. $AB = 6$ and $PB = 3\sqrt{3}$

34. $AD = \dfrac{8}{\sqrt{13}}$ and $PD = \dfrac{12}{\sqrt{13}}$

37. $\angle PTA = \angle PBT$; therefore triangles PTA and PBT are similar.

38. Use Ex. 37.

39. $\dfrac{20}{7}$

40. 7.4

41. $\dfrac{77}{6}$

42. Letting $CP = x$, we have $x^2 + 5x = 96$.

   We can find an approximate solution by trial-and-error. For example, 7 is too small, and 8 is too large; $7\frac{1}{2}$ seems about right; try it. We get $56 + \frac{1}{4} + 37 + \frac{1}{2} = 93\frac{3}{4}$, which is a bit small. So we try 7.6, getting $57.76 + 38.0 = 95.76$, and this is very close.

   Applying the quadratic formula to the equation $x^2 + 5x - 96 = 0$ yields $x = \dfrac{-5 + \sqrt{409}}{2} = 7.6$ and another value that we reject because it is negative.

43. $2\sqrt{15}$

44. Letting $AP = x$, we have $x^2 + 4x = 64$.

   We can find an approximate solution by trial-and-error. For example, 6 is too small and 7 is too large; $6\frac{1}{4}$ seems about right; try it. We get $(6 + \frac{1}{4})^2 + 4(\frac{25}{4}) = 36 + 3 + \frac{1}{16} + 25 = 64\frac{1}{16}$. So $6\frac{1}{4}$ is very close indeed.

   Applying the quadratic formula to the equation $x^2 + 4x - 64 = 0$ yields $x = 2\sqrt{17} - 2 = 6.24$ and another value that we reject because it is negative.

45. 56.25

47. See note on Ex. 14.

In Fig. 33, let A and B move toward each other along the minor arc
AB; then let C and D move toward each other along the minor arc CD.
In the case that both secants become tangents we have the situation
in Ex. 8 on page 141.

Pages 154-155: Exercises.

1. The first two paragraphs of the proof of Theorem 23 on page 153 can
be applied to any triangle ABC.

2. In Fig. 36, triangles OAB and OBC are equal isosceles triangles; so
∠ ABO = ∠ CBO.

3. If we regard Fig. 36 as representing part of an inscribed equilateral
polygon, the equality of the base angles of the several equal isosceles
triangles is sufficient to prove that ∠ A = ∠ B = ∠ C = ∠ D = ...
Thus the definition of <u>regular</u> <u>polygon</u> on page 85 is satisfied.

If the equilateral polygon is formed by joining every second, or
every third, or every fourth, ..., point of division on the circle,
then the polygon will be a star when n is odd.

4. See Figs. A and B.

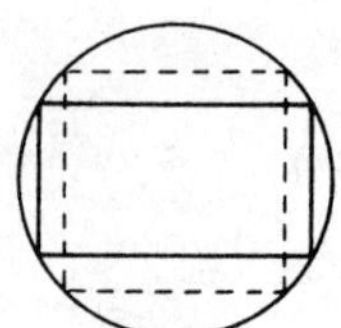
Fig. A

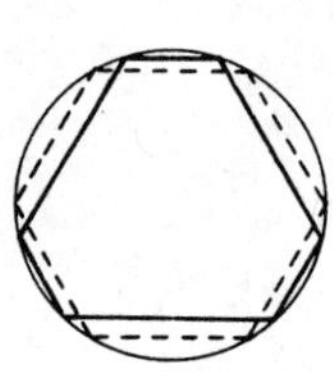
Fig. B

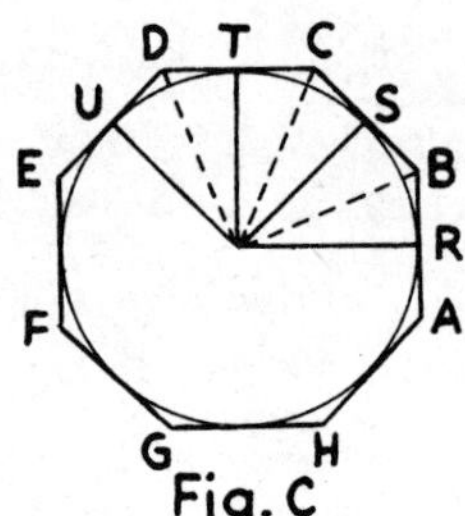
Fig. C

5. In Fig. C, polygon ABCDEFGH is equi-angular. The inscribed circle
touches the sides of the polygon at R, S, T, U, . . . , so that triangles ROB, BOS, SOC, COT, ... are similar. For the angles at R,
S, T, U, ... are right angles (Theorem 21) and the angles at B, C,
D, ... are halves of equal angles. These triangles are also equal,
since OR = OS = OT = OU = ... Therefore RB = BS = SC = CT =

TD = ..., and BC = CD = ..., so that ABCDEFGH is a regular

polygon.

6. See Figs. A and B below.

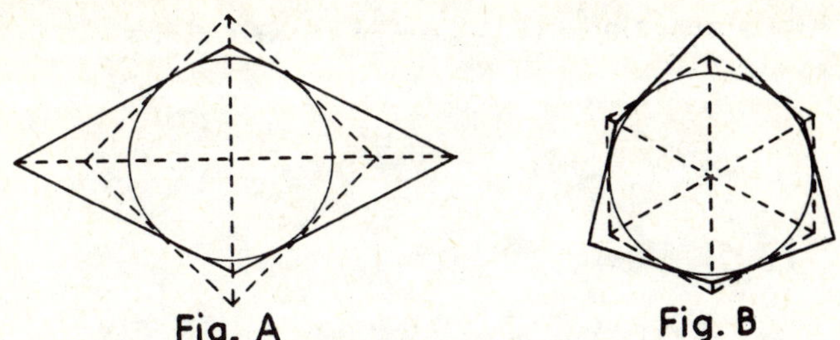

Fig. A   Fig. B

7. Since each angle of the polygon is measured by (see footnote on page

145) the same number of equal arcs, all the angles of the polygon are

equal. The sides are all equal also (Theorem 19).

In Ex. 7 and Ex. 8 the phrase "any number" means "any integer

greater than two."

8. If the chords are drawn also, as in Ex. 7, we have n isosceles tri-

angles. In each triangle the angle between tangent and chord is the

same, so that the triangles are similar isosceles triangles. There-

fore the angles at the vertices of the circumscribed polygon are all

equal. Since the chords are all equal, these isosceles triangles

are not only similar, but equal; so that the sides of the circum-

scribed polygon are all equal.

9. One method of proof follows the pattern of the proof in Ex. 8, show-

ing first that the isosceles triangles are all similar, and then

that they are all equal.

A second method is merely to apply the theorem in Ex. 7 on this

page.

That the number of sides is doubled is sufficiently obvious without

expecting the student to give a formal proof.

10. This can be proved either by drawing chords and considering the

isosceles triangles, as in Exs. 8 and 9, or by immediate application

of the theorem in Ex. 8 on this page.

11. Prove angles equal by Ex. 6 on page 85. Since AB = BC = CD = ...

and A'B' = B'C' = C'D' = ..., $\dfrac{AB}{A'B'} = \dfrac{BC}{B'C'} = \dfrac{CD}{C'D'} = $ . . . .

12. $\dfrac{5}{\sqrt{2}}$

13. $7\sqrt{2}$

14. $\dfrac{4}{\sqrt{3}}$

15. $7\sqrt{3}$

16. $30\sqrt{3}$

<u>Pages 157-161: Exercises.</u>

1. $\angle SRT = \tfrac{1}{2}\angle O$

   $\angle RST = \tfrac{1}{2}\angle O'$

   $\angle SRT + \angle RST = \tfrac{1}{2}(\angle O + \angle O') = \tfrac{1}{2}(180^\circ)$

   Therefore $\angle RTS = 90^\circ$.

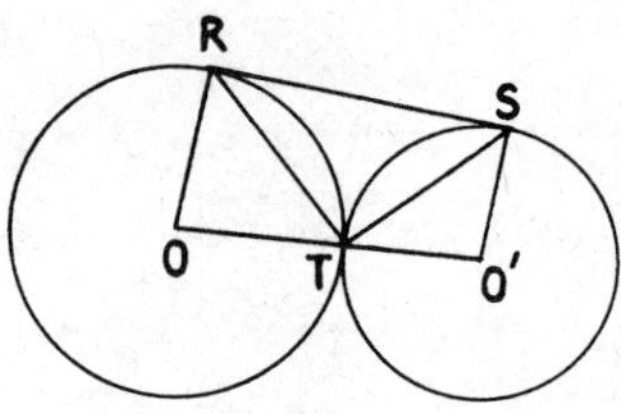

Fig. A

2. It is necessary only to prove that **the marked angles in each of the**

diagrams below are equal.

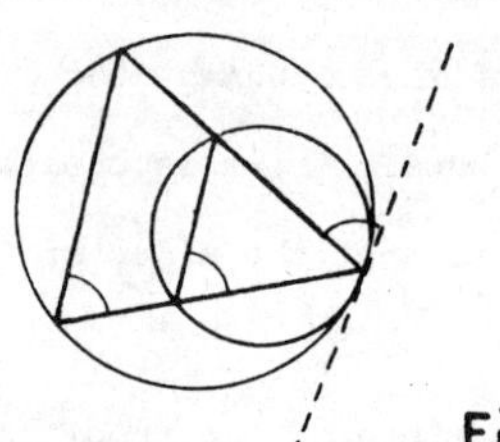
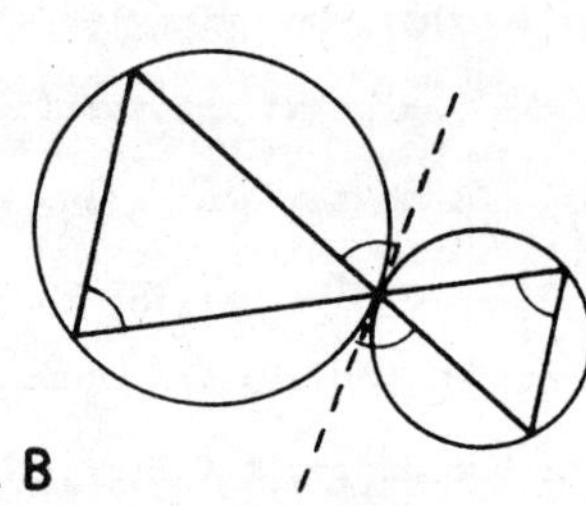
Fig. B

3. Use the preceding exercise and also Ex. 5 on page 127.

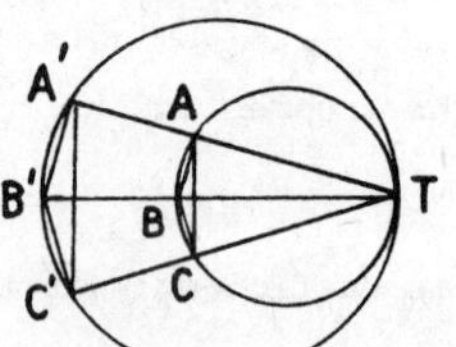

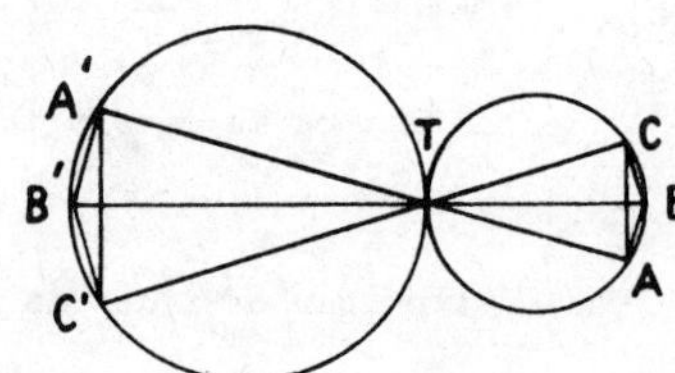

Fig. C

4. In Fig. A, TA' and TB' are random chords through T. It is sufficient
   to prove by Ex. 2 on this page that chords AB and A'B' are parallel.

5. See Fig. B. In each case $\angle$ ATB = 90°. Therefore $\angle$ A'TB' = 90° also,
   and A'B' is a diameter.

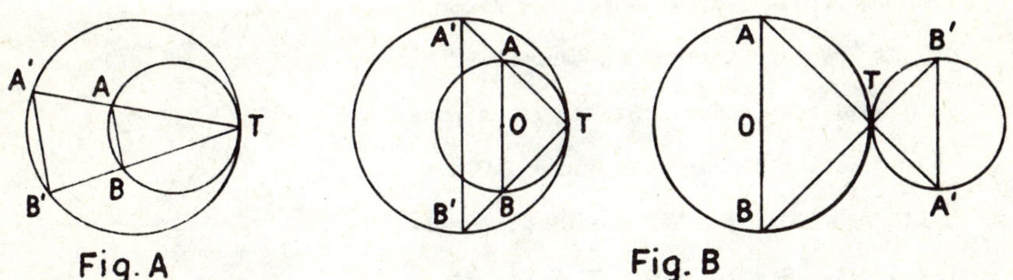

6. It is not necessary that all pupils suggest the same property. This
   is an interesting and significant diagram, and the total of all correct
   suggestions will enlighten everyone.

   See Fig. C below. P is equidistant from R, S, and T; Q is equidistant from U, V, and T.

   PQ = RS = UV. For PT = $\frac{1}{2}$RS; QT = $\frac{1}{2}$UV; and RS = UV because RM =
   UM and SM = VM.

7. Use the theorem in Ex. 37 on page 151.

8. See Fig. D. Triangle ADO is a 30° - 60° right triangle. Therefore
   DO = $\frac{1}{2}$AO = $\frac{1}{2}$CO, and DO = $\frac{1}{3}$DC.

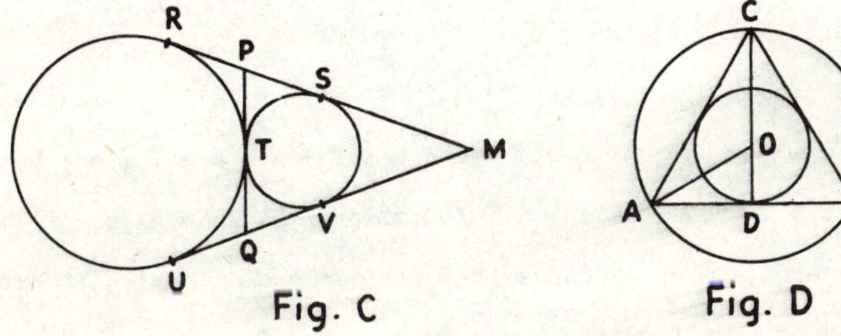

9. Since the inscribed angles C and D are measured by one half the same
arcs respectively, however CD may be drawn, the sizes of these two
angles do not vary. Consequently angle DBC does not vary in size.

10. The accompanying diagram shows circles
O and Q of Fig. 41, to which the lines
OA, QE, OQ, and the common tangent ST
at C have been added. By Ex. 16 on page
143, OQ passes through C. Since $\angle$ TCA =
$\angle$ SCE, $\frac{1}{2}\angle$ AOC = $\frac{1}{2}\angle$ EQC and OA is parallel to QE. Similarly, in Fig. 41, PA
is parallel to QD. But OAP is a straight line. Therefore EQD is
straight also.

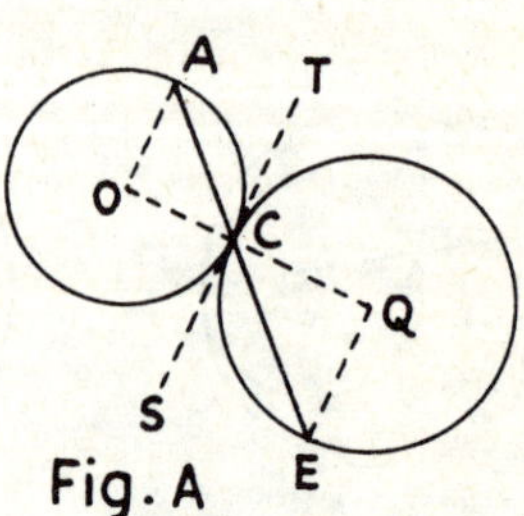

11. In Fig. 42 the angles at A and B are complementary. If the other
tangents from A and B are drawn, these angles at A and B will be
doubled; that is, they will be supplementary, and the new tangents
will be parallel.

12. The lengths of the segments in question are either the sum or the
difference of equal tangents from an external point.

13. If we letter the arcs a, b, c, d, e, ... in order, then in the
case of the equiangular polygon of five sides the equality of the
angles tells us that a + b + c = b + c + d = c + d + e = d + e + a =
e + a + b = a + b + c. It follows that a = d, b = e, c = a, d = b,
e = c, and hence that a = b = c = d = e. Since this polygon is both
equilateral and equiangular, it is regular.

    If the equiangular polygon has six sides, we get a + b + c + d =
b + c + d + e = c + d + e + f = d + e + f + a = e + f + a + b =
f + a + b + c = a + b + c + d. It follows that a = e, b = f, c = a,
d = b, e = c, f = d and hence that a = c = e and b = d = f; but there
is no way of equating a, c, or e to b, d, or f.

If the equiangular polygon has seven sides, we get a = f, b = g,
c = a, d = b, e = c, f = d, g = e. This is like the series of equations for the pentagon, except that in each equation we now skip four
letters instead of two. Since in each of these two cases the total
number of letters is odd, this skipping of an even number of letters
links all the letters and equations together. The same is true of any
polygon having an odd number of sides. Whereas in the case of any
polygon having an even number of sides, like the hexagon, we skip an
odd number of letters in each equation of the series; and since the
total number of letters is even, we succeed in linking only half the
letters into one series of equations, and the other half into a second series of equations; we can never bring the two series together.

14. In the case of the equilateral polygon of five
sides, or of any odd number of sides, the segments of the tangents are equal in pairs around
the polygon, as marked, beginning at vertex A,
until there is overlap; that is, until an $\underline{a}$
segment is seen to be equal to a $\underline{b}$ segment.
Consequently each vertex of the polygon is at

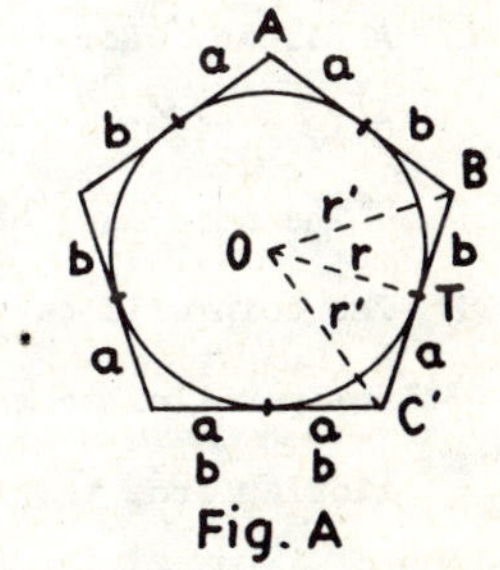

the same distance r' from the center of the given circle (Principle 12)
and the polygon can be inscribed in a circle of radius r'. By Ex. 3
on page 154 the polygon is regular.

Or, having proved an $\underline{a}$ segment equal to a $\underline{b}$ segment, we can prove
$\angle$ OBT equal to $\angle$ OCT, and hence $\angle$ B = $\angle$ C. In this way the equilateral polygon is proved to be equiangular also, and therefore regular.

In case the circumscribed equilateral polygon has an even number
of sides, the segments of the tangents are equal in pairs but without
overlap. It is impossible to prove that the polygon is regular.
Ex. 6 on page 155 affords examples of circumscribed equilateral

polygons that are not regular.

15. In Fig. 43, AO $\cdot$ OB = CO $\cdot$ OD;

$$AO \cdot OB = EO \cdot OP;$$

$$CO \cdot OD = EO \cdot OQ;$$

therefore OP = OQ.

Note: Exs. 16-23, like the other exercises on three-dimensional geometry in this book, are not meant to be logically connected with the two-dimensional geometry. They are included principally to challenge the pupil's imagination.

16. A circle. It is assumed, of course, that the pupil will think only of a right circular cone. To include a technical phrase of this sort in the question would add mystery rather than clarity.

17. A circle. Here also it is assumed that the pupil will consider only a right circular cylinder.

The coin must be held horizontal; parallel to the tilted cover. The coin will cast an elliptic shadow if the plane of the coin is not parallel to the floor (cover) and does not contain the perpendicular from the light to the floor (cover). In the latter case the shadow would be a "broad" line segment.

18. Circle. Meridian and equator are equal circles. Centers of all great circles are at the center of the sphere. Parallels of latitude are smaller circles than "great circles" and diminish as the latitude increases from equator to pole. The centers of all these circles are on the axis joining the two poles of the sphere.

19. The center of the sphere and the two given points on the surface of the sphere ordinarily determine a plane. This plane intersects the sphere in a great circle. Three points in a straight line do not "determine" a plane: this line can lie in a multitude of planes. This situation arises when the two given points are the extremities

of a diameter. For example, there are a multitude of meridians
through the north and south poles of the earth.

20. Cape Race, Newfoundland; Southern Ireland.

21. 75° in both cases

22. 5 hours. The sun's apparent motion around the earth covers 360
degrees in 24 hours; that is, 15 degrees in 1 hour.

23. One equilateral spherical triangle that is likely to occur to the
student is the triangle each of whose sides is a quadrant (90°) of a
great circle. If such a triangle be drawn on a tennis ball and then
an equilateral triangle with shorter sides - say 60° - be drawn inside the first triangle, the angles of the second triangle will
obviously be smaller than the 90° angles of the first triangle.
Pages 161-163: Exercises.

1. It does not follow; for consider the
triangles shown in Fig. A, in which
$JK = PQ$, $KL = QR$, and $\angle J = \angle P$.

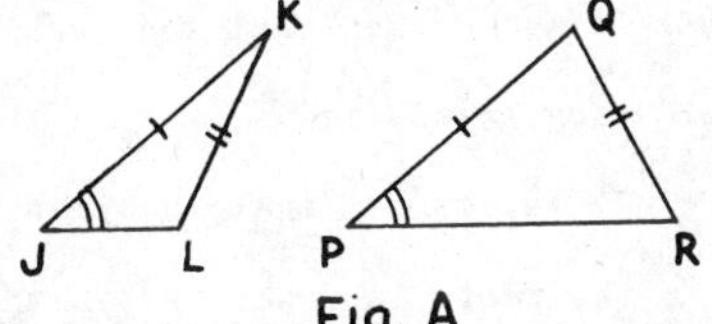

2. To say "draw FM parallel to CD"
demands too much. Either one should draw a line through F parallel
to CD and then prove that it meets BD at M; or else draw FM and prove
that FM is parallel to CD.

3. This is a case of "begging the question"; for the idea of equally
distant lines has no meaning for the student without the idea of
parallelism and so cannot be used in a definition of parallelism.

4. If it is reasonable to expect a team to win a majority of its games,
is it not equally reasonable that each of its opponents should expect
to win a majority of its games?

5. Are there any limits to the right of the taxpayers to see their
property?

6. It is necessary to know first how many heat units one ton of gas

works' coke yields. If it yields at least 9900 heat units per ton,
the Seacoal salesman's argument is worthless.

7. Or else the public school graduates do their college work more faith-
fully than do the private school graduates. There may be economic
and social reasons for this, quite apart from the earlier training
in school subjects.

8. Despite Blank's mistake, as he called it, he seems to have been very
successful in the world of business.

9. It is assumed that a self-emptying ash tray is an important enough
item to determine the choice of an automobile.

10. It is assumed that men's choices determine styles, rather than the
other way around.

11. It is assumed that the usual risk in a new business venture is not
very great.

12. It is assumed that the prosperity referred to was attributable to
the party in power rather than to economic forces that would have
been operating regardless of party politics.

13. It is assumed that those whose taxes remain unpaid for several years
are poor widows and the like.

<u>Page 168, line 5</u>: 1.4 inches

<u>Pages 169-170: Exercises.</u>

1. The length of CD is a trifle more than 5 centimeters and a trifle
less than 2 inches. It is easier to lay off 1 centimeter "plus a
hair" than to lay off $\frac{2}{5}$ of an inch "minus a hair." Nevertheless
with a scale of inches divided to sixteenths the student can approximate to $\frac{2}{5}$ of an inch. He must not collapse and quit because he hasn't
a scale divided precisely into fifths or tenths of inches. This question, therefore, is to some extent a test of the student's initiative
and resourcefulness.

2. The length of EF is a scant $2\frac{3}{4}$ inches, or 6.96 centimeters. The latter is more easily divided by 6. Indeed the decision to call the length 6.96 centimeters rather than 6.95 or 6.97 is influenced by the desired divisibility by 6.

3. 3.737, 4.010, 4.283

4. If $\frac{1}{m} = \frac{3}{2}$, merely use the third point of division in the answer to Ex. 1.

5. Using millimeters, $r:s:t = 36:28:20 = 9:7:5$. So we need to lay off $\frac{3}{7}$, $\frac{1}{3}$, $\frac{5}{21}$ of 7 centimeters, scant.

6. $\frac{14}{9}$ centimeters

7. 5.0(4) centimeters

8. $\angle AOB = 63^{\circ}$. Each third is 21°.

9. $\angle COD = 94^{\circ}$; $\angle EOF = 29^{\circ}$. The two parts are approximately 48° and 15°

10. About $43\frac{1}{2}^{\circ}$

11-12. Use method described on page 168.

Pages 171-172: Exercises.

1. Make angles of 135° at each end of AB and lay off lengths BC and AH equal to AB; and so on around.

2. Make angles of 140° at each end of AB and lay off lengths BC and AI equal to AB; and so on around.

3-4. Central angles must be 45° and 40° respectively.

Page 176, line 10. In any pair of triangles Principle 5 establishes pairs of corresponding angles equal. These equal angles establish the parallelism, by Theorem 14.

Page 176, lines 14-16. Draw parallels to BS through P, Q, and R. Proof of the construction depends on Theorem 16.

Page 178, line 8. By Ex. 21 on page 115.

Page 178, line 12. Proof depends on Ex. 20 on page 115.

Page 178, line 18. Proof depends on Principle 8.

1. Construct the perpendicular bisector of the chord of the arc and find
where it intersects the arc.

2. Draw the perpendicular through E to AB and find where the perpendicular intersects the diagonal AC.

3. $\sqrt{7}$ is the diagonal of a
square of side $\sqrt{3.5}$.

    At this point the teacher
may wish to show the class the
accompanying construction for $\sqrt{k}$:

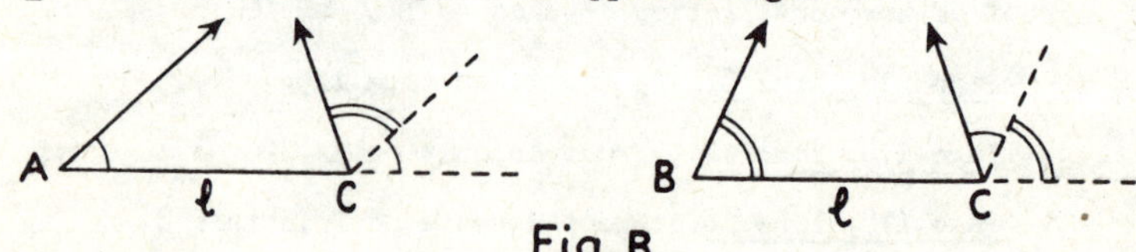

Fig. A

4. The unit of measure is not
specified, since it makes no difference what the unit is.

5. Fig. 28 involves three equilateral triangles. Fig. 29 involves a triangle having one angle equal to $60°$ and the sides including this angle in the ratio 2:1.   Fig. 30 involves an angle inscribed in a
semicircle. See Corollary 22c, page 146.

Page 185, line 10. OP and O'R are both perpendicular to the required
common internal tangent PR. A parallel through O' to PR will be perpendicular to OP extended and will meet it at M. O'M will then be tangent
at M to the circle having O as center and radius equal to OP + PM, or r + r'

    Page 186, line 22: Case 2. At one end of a line segment of length $\underline{1}$
construct an angle equal to $\angle A$ or to $\angle B$. At the other end of the line
segment construct an angle equal to $\angle C$ by first constructing an exterior
angle equal to $\angle A + \angle B$. See Fig. B. There are two cases, according as
side $\underline{1}$ is given opposite angle B or opposite angle A.

Fig. B

<u>Page 188, line 3</u>. The fifth and sixth situations in Fig. 46 on page 187.

<u>Pages 195-196: Exercises</u>.

1. Construct tangents to the circle at the vertices of an inscribed regular hexagon and extend these tangents until they meet (the bisectors of the central angles of the hexagon).

3. Trisect a right angle and bisect one of the $30°$ angles.

4. Inscribe a regular polygon of 15 sides and bisect a central angle. Then bisect again.

5. Inscribe a regular pentagon. The radius drawn to a vertex makes an angle of $54°$ with each adjacent side of the pentagon.

6. $108°$

7. Construct an angle of $108°$ by drawing a circle of any radius and inscribing a regular pentagon. Then, at each end of the given side AB construct an angle equal to $108°$. And so on around.

8. At each end of the given side AB construct an angle of $135°$, presumably by erecting perpendiculars at A and B and bisecting the right angle between each perpendicular and AB extended.

9. As in Ex. 7.

10. From one vertex of the given polygon draw n-3 diagonals and copy the appropriate angles. This is easier than using the construction for the fourth proportional to three given line segments.

11. In equilateral triangle ABC (Fig. A), construct the three medians, meeting in O. The bisector of angle BFO will meet BO at a point G that is equidistant from BD, BF, DO, and FO. Therefore G is the center of one of the desired circles. Another and much harder method is to mark off on BO the distance BG equal to $\dfrac{AB}{\sqrt{3}+1}$.

12. Draw the diagonals of the square. The midpoint of each half of a diagonal will be the center of one of the desired circles.

13. In the given square ABCD (see Fig. A on this page) draw the diagonals AC and BD, meeting at O. Construct the bisectors of angles OAB and OBA. These bisectors meet at Q, the center of one of the desired circles.

A slightly easier construction, but somewhat harder to justify, is to draw the diagonals AC and BD, meeting at O; draw arcs with centers C and D and radii CO and DO respectively to determine the points E, F, G, and H and thus determine the intersection R of EF and GH. R is the center of one of the desired circles. For if the side of the given square be $\underline{s}$ and if the radius of one of the desired circles be $\underline{r}$, then Fig. B shows that $\frac{s}{2} = r + r\sqrt{2}$. But Fig. B also shows that if $\underline{t}$ be one side of the regular octagon GFIH - - -, then $\frac{s}{2} = \frac{t}{2} + \frac{t}{\sqrt{2}} = \frac{t}{2}(1 + \sqrt{2})$. Consequently $r = \frac{t}{2}$ and we can utilize the regular octagon to locate the centers of the desired circles.

It is clear from Fig. A that $DO = DG = DH = \frac{s}{\sqrt{2}}$ and that $DG + FC = s + FG$. That is, $2\left(\frac{s}{\sqrt{2}}\right) = s + FG$ and $FG = s(\sqrt{2} - 1)$. But $s = t + 2\left(\frac{t}{\sqrt{2}}\right) = t(1 + \sqrt{2})$, and $t = \frac{s}{\sqrt{2} + 1} = s(\sqrt{2} - 1)$. Therefore $FG = t$.

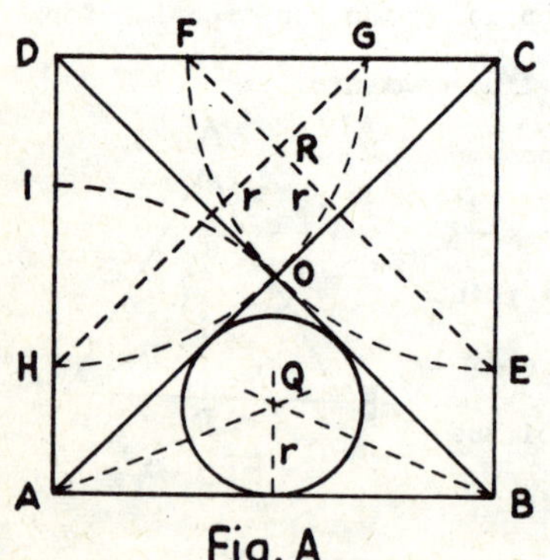

Fig. A

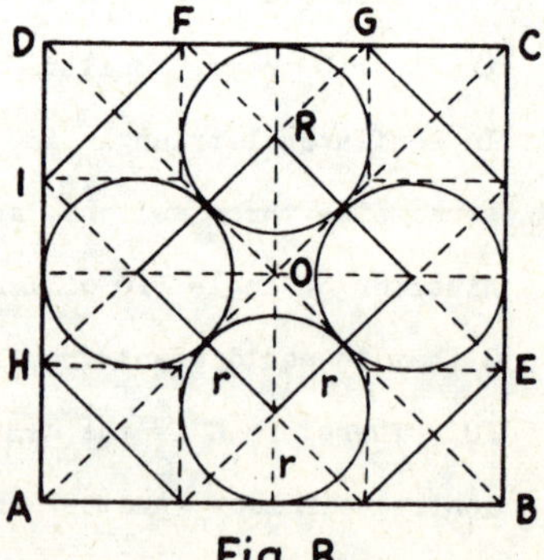

Fig. B

15. The single-marked flaps should be pasted first, then the double-
    marked flaps. The face marked L will be the last to be stuck down.
    See Fig. A.

16. First draw a random circle, inscribe a regular pentagon ABCDE, and
    draw diagonal AC. Then at each end of the given line-segment A'C'
    construct an angle equal to angle BAC and thus determine B'.

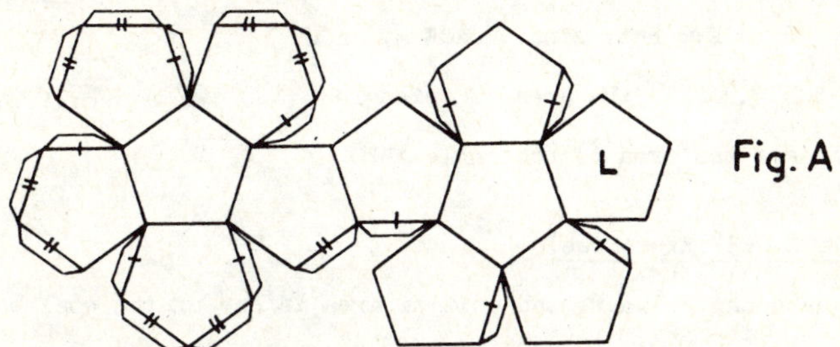

    If instead of using angles one wishes to use lengths, it is neces-
sary to note that all the acute-angled triangles formed by the sides
and diagonals of a regular pentagon are isosceles triangles with
angles 72°-36°-72°. In
every triangle of this
sort, such as triangle
FBG in Fig. B, the short
side is to one of the
longer sides as $\sqrt{5} - 1$ is
to 2. (See page 192). If

we take FG as $\sqrt{5} - 1$ and BF as 2, then AF = GC = 2; AB = AG = $\sqrt{5} + 1$;
and AC = $\sqrt{5} + 3$. So the ratio of the desired length AB to the given
length AC is as $\sqrt{5} + 1$ is to $\sqrt{5} + 3$.

17. If r be the radius of the given circle, draw a circle with center O
    that shall have a radius equal to $\sqrt{r^2 + 1^2}$. This circle will cut
    the given line in two points from either of which tangents to the
    given circle will be of length 1.

<u>Page 199: Exercises.</u>

1. 137 square units

<u>Page 200: Exercises.</u>

1. In the similar triangles ABC and CBE,

   $\frac{AC}{AB} = \frac{CE}{BC}$. Therefore $\frac{1}{2}$AC x BC = $\frac{1}{2}$AB x CE.

2. See Fig. A. The area of a rectangle

   ABFG = AB x CE. But, since $\triangle$ ACE = $\triangle$ ACG

   and $\triangle$ BCE = $\triangle$ BCF, the area of triangle

   ABC is half the area of rectangle ABFG.

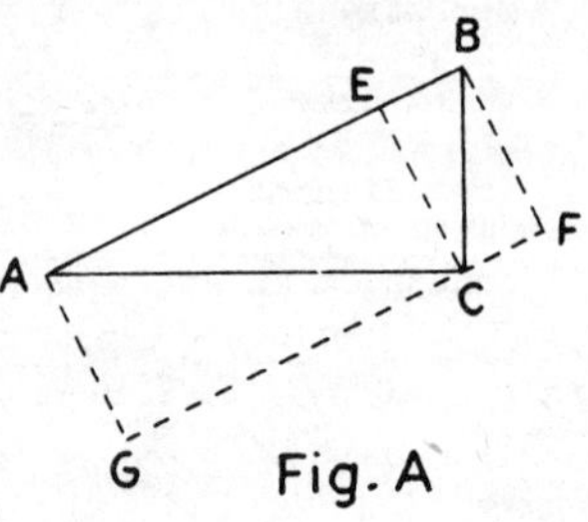

<u>Pages 202-203: Exercises.</u>

1. 905 square units, where the unit of area is one of the smallest

   squares of the squared paper. This can be found either by assigning

   coordinates to each vertex of the diagram as in Ex. 1 on page 199, or

   by counting the number of squares inside the boundary.

2-3. The area of the left-hand triangle in Fig. 3 is $\frac{1}{2}$ x $\frac{26}{16}$ in. x $\frac{8\frac{1}{2}}{16}$ in. =

   $\frac{1}{2}$ x $\frac{221}{256}$ sq. in., using $\frac{1}{2}$AC x BD; or $\frac{1}{2}$ x $\frac{20\frac{1}{2}}{16}$ in. x $\frac{11}{16}$ in. = $\frac{1}{2}$ x $\frac{225\frac{1}{2}}{256}$ sq.

   in., using $\frac{1}{2}$AB times the altitude from C. In either case the area is

   approximately $\frac{7}{16}$ of a square inch. The area of the right-hand tri-

   angle is $\frac{1}{2}$ x $\frac{18\frac{1}{2}}{16}$ in. x $\frac{1}{2}$ in. = $\frac{37}{128}$ sq. in., using $\frac{1}{2}$AC x BD; or

   $\frac{1}{2}$ x $\frac{26}{16}$ in. x $\frac{3}{8}$ in. = $\frac{39}{128}$ sq. in., using $\frac{1}{2}$AB times the altitude from C.

   In either case the area is approximately $\frac{19}{64}$ of a square inch.

4. 29 millimeters x 17 millimeters = 493 sq. mm., or $\frac{18}{16}$ in. x $\frac{10\frac{1}{2}}{16}$ in. =

   $\frac{189}{256}$ sq. in. = 0.738 sq. in.

5. Area equals $\frac{\frac{1}{2}(29 + 10\frac{1}{2})}{16}$ in. x $\frac{1}{2}$ in. = $\frac{79}{128}$ sq. in., or about $\frac{5}{8}$ of

   a square inch.

6. $\frac{s^2}{4}\sqrt{3}$

7. $e^2\sqrt{3}$

8. Add the area of the three parallelogram faces to the area of both

triangular bases. If the prism is a right prism, the three faces are rectangles, and their area is equal to the altitude of the prism times the perimeter of one of the triangular bases.

9. $\dfrac{\frac{1}{2}bh}{\frac{1}{2}b'h'} = \dfrac{b \cdot b}{b' \cdot b'}$, since $\dfrac{h}{h'} = \dfrac{a}{a'} = \dfrac{b}{b'} = \dfrac{c}{c'}$

10. $\dfrac{\triangle ABC}{\triangle ADC} = \dfrac{\frac{1}{2}AB \times h}{\frac{1}{2}AD \times h} = \dfrac{AB}{AD}$; also $\dfrac{\triangle ADC}{\triangle ADE} = \dfrac{\frac{1}{2}AC \times h'}{\frac{1}{2}AE \times h'} = \dfrac{AC}{AE}$. Multiplying,

we get $\dfrac{\triangle ABC}{\triangle ADE} = \dfrac{AB \times AC}{AD \times AE}$.

<u>Pages 204-205: Exercises.</u>

1. Add the areas of the isosceles triangles in Fig. 11.

2. Apothem = $\frac{5}{2}\sqrt{3}$ inches; area = $\frac{75}{2}\sqrt{3}$ square inches

3. Area = $\frac{3}{2}\sqrt{3}\, r^2$

4. Perimeter = $3\sqrt{3}\, r$; area = $\frac{3}{4}\sqrt{3}\, r^2$

5. Apothem = $\dfrac{s}{2\sqrt{3}}$; radius = $\dfrac{s}{\sqrt{3}}$

6. On page 193 the apothem $r - y$ is shown to be $r\left(\dfrac{1 + \sqrt{5}}{4}\right)$. Therefore

the area is $\dfrac{5}{2} \cdot \dfrac{r}{2}\sqrt{10 - 2\sqrt{5}} \cdot \dfrac{r}{4}(1 + \sqrt{5})$. This equals

$\dfrac{5}{16}\, r^2 \sqrt{(10 - 2\sqrt{5})(6 + 2\sqrt{5})}$, or $\dfrac{5}{8}\, r^2 \sqrt{10 + 2\sqrt{5}}$.

7. Using the measurements shown in the accompanying diagrams, the area of the left-hand polygon is 455 sq. mm. and the area of the right-hand polygon is 456 sq. mm. The numbers inside the triangles are altitudes.

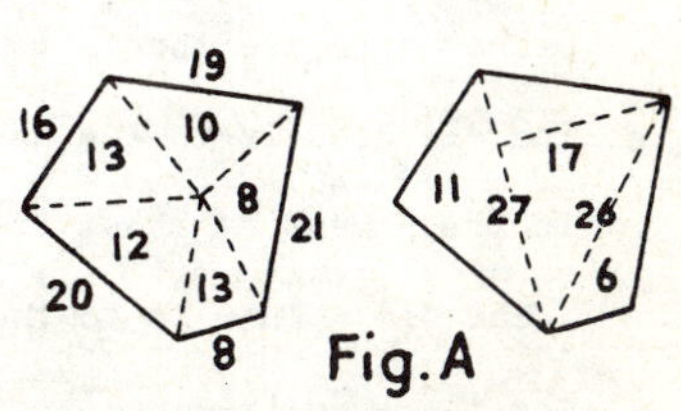

8. Counting <u>up</u>, we have $20 + 108 + 120 + 110 + 93 = 451$ sq. mm.

9. Counting from left to right we have $48 + 105 + 103 + 65 + 40 + 49 + 70 + 61 + 2 = 543$ sq. mm.

10. If the lengths of the sides of the given polygons are s_1 and s_2

respectively, then $\dfrac{p_1}{p_2} = \dfrac{s_1}{s_2} = \dfrac{r_1}{r_2}$.

11. From Ex. 1 and Ex. 10 we know that $\dfrac{A_1}{A_2} = \dfrac{\frac{1}{2}p_1 a_1}{\frac{1}{2}p_2 a_2} = \dfrac{r_1 a_1}{r_2 a_2}$

Therefore $\dfrac{A_1}{A_2} = \dfrac{r_1^{\,2}}{r_2^{\,2}}$.

Pages 207-208: Exercises.

1. 1 to 400

2. $\dfrac{81}{64}$

3. It is assumed in this exercise, of course, that the three sides of the right triangle are corresponding sides of the three similar polygons. If the lengths of the three sides of the triangle, arranged in order of increasing magnitude, are $\underline{a}$, $\underline{b}$, $\underline{c}$, and if the corresponding polygons are designated I, II, III, then, by Theorem 27, $\dfrac{I}{II} = \dfrac{a^2}{b^2}$ and $\dfrac{I}{III} = \dfrac{a^2}{c^2}$. This means that if area I is equal to $m \cdot a^2$, then area II is $m \cdot b^2$, and area III is $m \cdot c^2$. But $a^2 + b^2 = c^2$, by Principle 12. It follows that $ma^2 + mb^2 = mc^2$ and that I + II = III.

4. An edge of the new cube must be $\sqrt[3]{2}$ times the edge of the given cube.

5. The area of the new cube must be $\sqrt[3]{4}$ times the area of the given cube.

6. The ratio of the volumes of the two sizes of cup is 64:125. So two 5-cent cups are the better buy.

7. The old area is to the new area as 1600 is to 3025. So the increase in the amount of sheet iron is $\dfrac{1425}{1600}$ of the old amount, or 89.1%.

   The old volume is to the new volume as 64,000 is to 166,375. So the increase in capacity is $\dfrac{102,375}{64,000}$ of the old capacity, or just a trifle under 160%.

8. The sum of the squares on the other two sides of the right triangle is equal to the outside square minus two rectangles; and these two rectangles are equal to four right triangles.

9. The square on the hypotenuse is equal to the inner tilted square

plus four right triangles. The sum of the squares on the other two
sides is equal also to the inner tilted square plus four right triangles.

Pages 213-217: Exercises.

1. 45.5 in.; 58.6(4) feet or 58 ft., 8 in.; 22 cm.

2. 165 sq. in.; 274 sq. ft.; 38.5 sq. cm.

3. 1.138 in.; 14.4 cm.

4. 2.489 ft.; 9.46 cm.

5. 18.4 sq. ft.; 58 sq. in.; $\dfrac{c^2}{4\pi}$ sq. in.

6. $c = 2\sqrt{\pi}\,\sqrt{A} = 3.54\sqrt{A}$.   87.1 cm.; 15.4 ft.; 32 in.

7. The two central angles in the triangles in Fig. 23 are equal, by
Principle 8. Therefore each of the corresponding arcs is the same
fractional part of its circumference. Since the circumferences have
the same ratio as their radii, the arcs do also.

8. $\dfrac{A_1}{A_2} = \dfrac{r_1^2}{r_2^2} = \dfrac{c_1^2}{c_2^2}$

9. 4

10. 50 sq. in.

11. $2\frac{1}{4}$

12. $\dfrac{3}{2\sqrt{2}}$ or $\frac{3}{4}\sqrt{2}$

13. $\dfrac{15^2}{16^2}$ x 20 = 17.6 sq. in.

14. 4

15. $\sqrt{5}$, or 2.236

16. $4r^2 - \pi r^2 = \frac{6}{7} r^2$; $\pi r^2 - 2r^2 = \frac{8}{7} r^2$.

17. $\left(\dfrac{22}{7} \times 3 \times 7\right) + 2\left(\dfrac{22}{7} \cdot \dfrac{9}{4}\right) = 80\frac{1}{7}$ sq. in.

18. 7.6 inches

19. 4π

20. $\dfrac{4}{3}\pi$

21. $\dfrac{64}{3}\pi + 32\sqrt{3}$

22. Since $c^2 = a^2 + b^2$, $\frac{\pi}{2} \cdot \frac{c^2}{4} = \frac{\pi}{2} \cdot \frac{a^2}{4} + \frac{\pi}{2} \cdot \frac{b^2}{4}$. That is, the area of the largest semicircle is equal to the sum of the areas of the other two semicircles.

Both the area of the triangle and the sum of the areas of the two shaded figures are equal to the area of the largest semicircle minus the areas of two circular segments. (The term "circular segment" is not used in the text but will be clear to teachers at this point.)

Professor Norman Anning of the University of Michigan suggests that it is pertinent for teachers to point out that the Greeks, in their search for a means of computing the area of a circle, believed they were on the way to success when they could compute the area of a figure bounded entirely by arcs of circles. We now know that success was not to be attained in this way.

23. On page 211, $p_8 = 6.1232 \times 12$. Therefore $s_8 = 9.18$.

24. On page 211, $s_{2n} = \sqrt{2r^2 - r\sqrt{4r^2 - (s_n)^2}}$. When n = 6 we have $s_{12} = \sqrt{2r^2 - r^2\sqrt{3}} = r\sqrt{2 - \sqrt{3}} = .518r$. Therefore $p_{12} = 6.216r$, which falls short of the circumference by a little less than $0.07r$.

25. Perimeter = $14 \times 5 \sin\left(\frac{360^\circ}{14}\right) = 70 \sin 25.7^\circ = 30.36$

26. The radius of the silo is about 8.3 feet. The sine of half the angle at the vertex is approximately $\frac{8.3}{11.2}$, or 0.741. Therefore the angle at the vertex is about 96°.

27. The term "lateral area" is probably new to the pupil, but clear enough from the context. The pupil has merely to add the areas of all the lateral faces of the prism. The theorem is not true unless all the lateral faces of the prism are perpendicular to each base; that is, unless the prism is a right prism. The bases need not be regular polygons.

28. The pupil can think of the cylindrical surface as slit parallel to the axis, unfolded, and laid out flat. The bases of the cylinder

need not be circles, but the axis of the cylinder must be perpendicular to each base.

29. The term "slant height" should be clear from the formula and Fig. 30.

30. If the pupil will think of the conical surface as slit along an
"element" of the cone, unfolded, and laid out flat, he will see that

he is asked to find the area of a sector of a circle of radius l and
of arc length c. The formula for the area of a sector, $\frac{1}{2}rs$, given
on page 213, becomes in this case $\frac{1}{2}lc$, or πrl.

Another way of regarding the lateral area of the cone is as the
limit of the lateral area of circumscribed regular pyramids as the
number of faces is indefinitely increased. That is, the limit of
$\frac{1}{2}pl$ is $\frac{1}{2}cl$, or πrl.

31. Extend one of the sides of the given polygon to form an exterior
angle. Reproduce this angle at the center of the given circle, thus
determining two vertices of the new polygon.

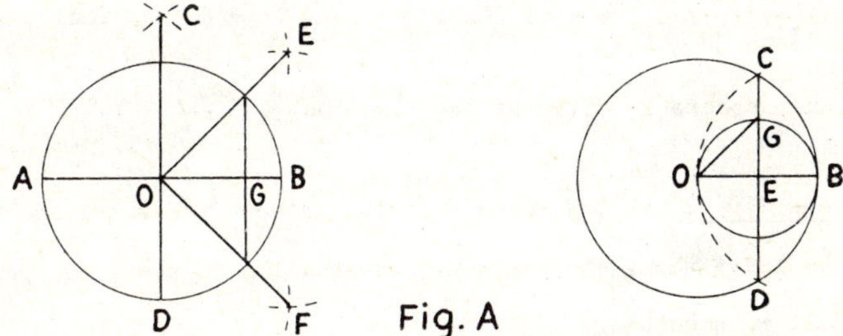

Fig. A

32. (See note following Ex. 36) If the radius of the given circle is r,
the radius of the inner circle will be $\frac{r}{\sqrt{2}}$. This length is OG in
each of the suggested constructions shown in Fig. A, in which all
the points except O are determined in alphabetical order.

33. Determine r_1 and r_2 so that $r_1:r_2:r = 1:\sqrt{2}:\sqrt{3}$. That is, $r_1 = \frac{r}{\sqrt{3}}$
and $r_2 = \sqrt{2}\,r_1 = \frac{\sqrt{2}}{\sqrt{3}}r$. Fig. A on the next page, in which the points

are determined in alphabetical order, shows one way of
constructing AG equal to $\dfrac{r}{\sqrt{3}}$. AG times $\sqrt{2}$
will give r_2.

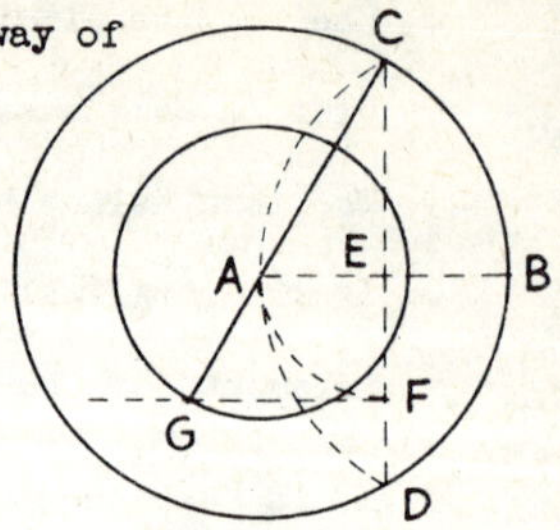

34. Since $r_1 : r_2 : r_3 : \ldots : r = 1 : \sqrt{2} : \sqrt{3} : \ldots : \sqrt{n}$,
the radius r_k of the kth inner circle is
given by the equation $r_k = \dfrac{\sqrt{k}}{\sqrt{n}} \cdot r$.

35. Multiply any side of the given polygon by $\sqrt{n}$
to get the corresponding side
of the desired polygon.

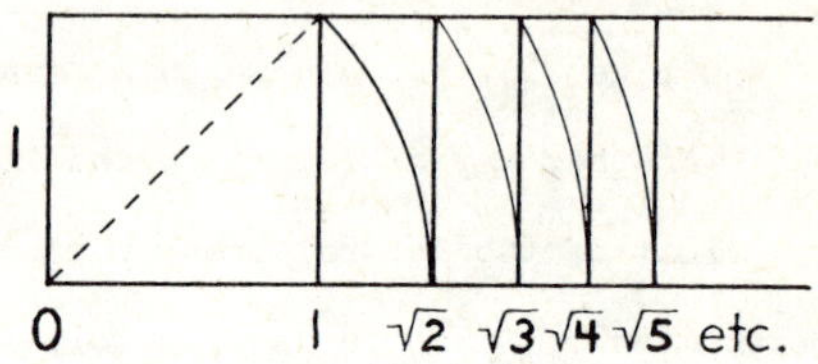

36. Multiply the radius of the
given circle by $\sqrt{\dfrac{m}{n}}$.

   Note: Exercises 32-36 can be
solved also by means of the
diagram at the right.

37. Their volumes have the ratio 1 to 8.

38. Their radii have the ratio 1 to $\sqrt[3]{2}$.

39. Since $r_1 : r_2 : r_3 \ldots : r = 1 : \sqrt[3]{2} : \sqrt[3]{3} : \ldots : \sqrt[3]{n}$, the radius r_k of the
k^{th} inner sphere is given by the equation $r_k = \dfrac{\sqrt[3]{k}}{\sqrt[3]{n}} \cdot r$.

   <u>Page 219, lines 5-7</u>. The friction between the water and the pipe
will be least when, for a given cross-section of area, the perimeter is
as small as possible.

   <u>Page 219, lines 8-9</u>. Pinching the outer end of the exhaust pipe re-
duces the area of cross-section of the pipe. This increases the veloci-
ty of the exhaust gases, interfering with the vibration in such a way as
to reduce the noise of the exhaust. (It is not expected that pupils will
be able to answer this question from their knowledge of geometry alone.
It is expected that they will ask someone who knows something about
automobile engines.)

Page 219, lines 12-14. Since the spherical one has less surface it will be less exposed to the dissolving effect of the saliva.

Page 220, lines 1-3. Four? Yes, the regular octahedron. Five? Yes, the regular icosahedron. See Page 189. Six? No; for if so, then six faces would have a common vertex with six 60° angles at that point and the corner would be flattened down until the vertex ceased to exist.

Page 220, lines 4-6. Three regular pentagons? Yes, the regular do-decahedron, page 196. More than three? No. Three regular hexagons? No.

Page 220, lines 7-10. The five convex regular polyhedra are:

1. The <u>regular</u> <u>tetrahedron</u>, four faces, in which three equilateral triangles meet at each vertex. See page 188.

2. The <u>cube</u>, six faces, in which three squares meet at each vertex.

3. The <u>regular</u> <u>octahedron</u>, eight faces, in which four equilateral triangles meet at each vertex.

4. The <u>regular</u> <u>dodecahedron</u>, twelve faces, in which three regular pentagons meet at each vertex.

5. The <u>regular</u> <u>icosahedron</u>, twenty faces, in which five equilateral triangles meet at each vertex.

Page 221: Review Exercises.

1. Through each midpoint draw a line parallel to the line through the other two midpoints.

2. In triangles BCD and CBE (Fig. A) two angles of one are equal respectively to two angles of the other. Therefore these triangles, having BC in common, are equal, and CD = BE. Since ED

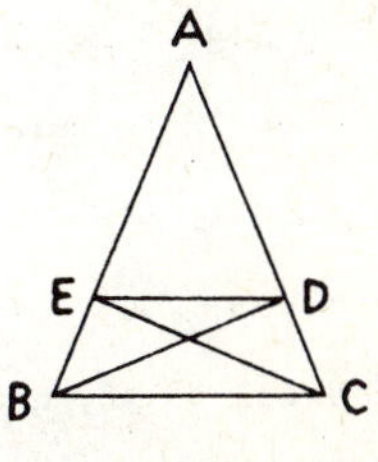

Fig. A

divides AB and AC proportionally, it is parallel to BC. (By Ex. 21 on page 115, or directly by means of Principle 5, Case 1 of Similarity.)

3. Make a regular five-pointed star by extending the sides of a regular

pentagon, as shown in Fig. A. Each angle of the pentagon is 108°;
each interior angle is 72°; the angle at each point of the star is 36°.

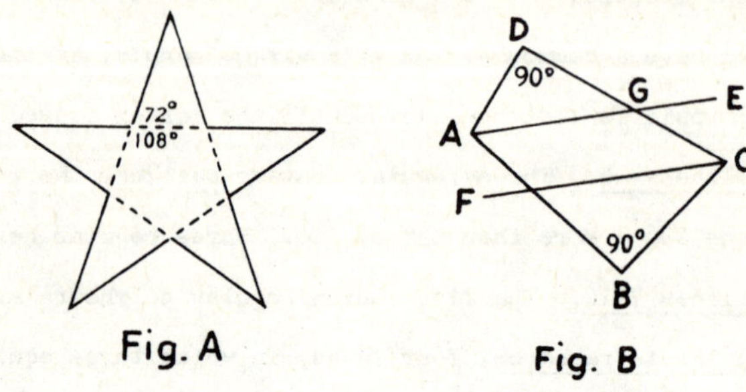

4. Since in Fig. B $\angle B = \angle D = 90°$, $\angle A + \angle C = 180°$, and $\angle A$ must be
less than 180°. Consequently AE, the bisector of angle A, must meet
side DC at a point G which will be between D and C, or at C, or beyond
C. In any case, $\angle AGD = 90° - \frac{1}{2}A$ by Principle 9. But $\angle BCF = 90° -
\frac{1}{2}A$ also, since $\angle A + \angle C = 180°$. Therefore the bisectors AE and CF
meet DC at the same angle and so are either parallel or coincident,
by Theorem 14.

5. In Fig. C let H be the mid-point of AD.
Then HM is parallel to DC, and consequently also to AB. Similarly HN, not
assumed to contain the point M, is parallel to AB, and consequently also to

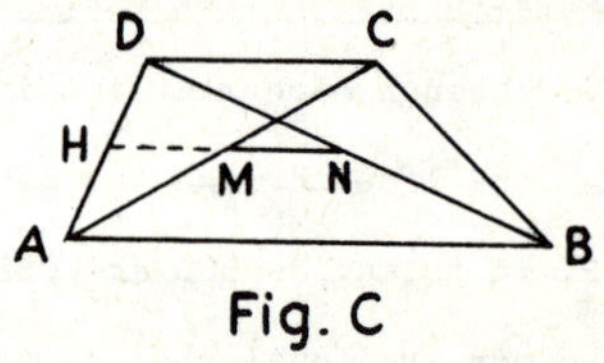

DC. Therefore HM and HN must coincide (by Theorem 13), and MN is
parallel to AB and CD.

6. Following the sort of argument used in Ex. 4 on page 127, let the
line joining the mid-points M and N in Fig. A on the next page meet
AD extended at P and meet BC extended at Q.

Then $\dfrac{PM}{PN} = \dfrac{AM}{DN} = \dfrac{MB}{NC} = \dfrac{QM}{QN}$,

$\dfrac{PM}{PN} - 1 = \dfrac{QM}{QN} - 1$,

$\dfrac{NM}{PN} = \dfrac{NM}{QN}$, and $PN = QN$.

Therefore P and Q must coincide at the point

R which is common to AD extended and BC extended.

7. In Fig. B let the line joining the midpoints

of M and N meet AC at P and meet BD at Q.

Then $\dfrac{PM}{PN} = \dfrac{AM}{CN} = \dfrac{BM}{DN} = \dfrac{QM}{QN}$,

$\dfrac{PM}{PN} + 1 = \dfrac{QM}{QN} + 1$,

$\dfrac{MN}{PN} = \dfrac{MN}{QN}$, and $PN = QN$.

Therefore P and Q must coincide

at point S which is common to both AC and BD.

First alternative proof:

Make the proof depend on Ex. 4, page 127, which states that if

three lines cut off proportional segments on two parallel lines, they

are either parallel or concurrent. In this case two of the three

lines are diagonals of the trapezoid and hence must intersect. Con-

sequently all three lines AC, BD, and MN, must intersect.

Second alternative proof:

In Fig. C the diagonals AC and BD

intersect at O. Lines OM and ON

join O to the midpoints of AB and

CD. We must prove that MON is a

straight line.

By Theorem 15 and Case 2 of Similarity, triangle OAB is similar to

triangle OCD. Therefore $OC = k \cdot OA$, $CD = k \cdot AB$; and $CN = k \cdot AM$. By

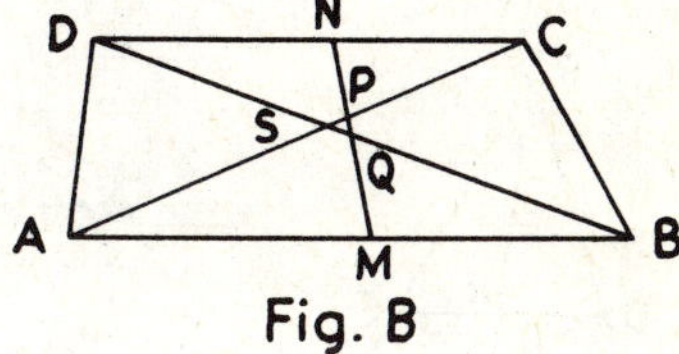

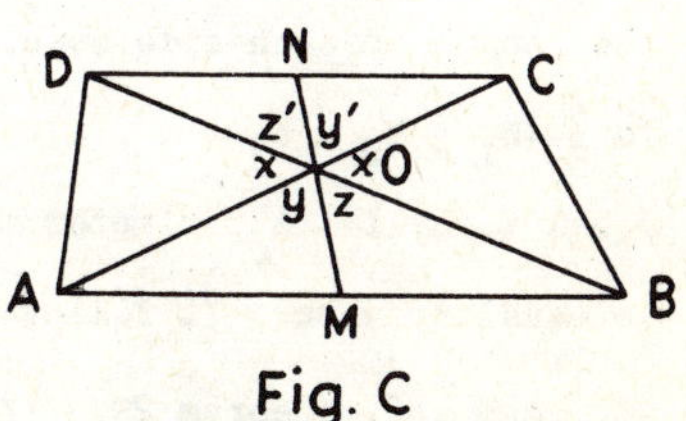

Case 1 of Similarity, triangles OAM and OCN are similar and triangles OMB and OND are similar. Consequently $\angle y = \angle y'$, $\angle z = \angle z'$; $\angle x + \angle y + \angle z' = \angle x + \angle y' + \angle z = 180^\circ$; and MON is a straight line.

8. The line joining the midpoints of the bases of a trapezoid passes through the point of intersection of the diagonals and through the point of intersection of the non-parallel sides extended. See Fig. A.

9. In Fig. B, arc AB = arc CD (by Theorem 19) and arc AC = arc BD (by Ex. 3, page 141). Therefore arc AB + arc AC = arc CD + arc BD = 180°.

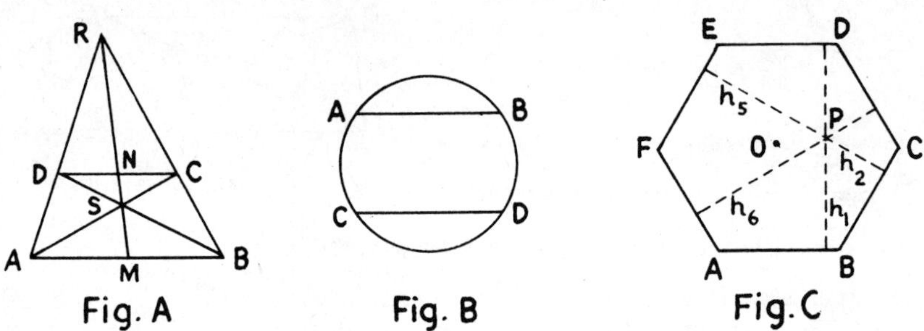

10. From any point P within the regular polygon draw lines to the vertices A, B, C, . . . and draw perpendiculars to the sides, extending the sides if necessary. Let the lengths of these perpendiculars from P to AB, BC, CD, ... be called h_1, h_2, h_3, ... respectively. Let the length of each side be <u>s</u>. Then the area of the polygon is equal to $\frac{s}{2}$ ($h_1 + h_2 + h_3 + \dots$). But the area of the polygon is also equal to half the perimeter times the apothem, a; that is, the area is equal to $\frac{1}{2}nsa$. It follows that ($h_1 + h_2 + h_3 + \dots$) = na.

<u>Page 232, Theorem 29.</u> If a is not greater than b, then either a = b or a < b. But each of these alternatives has a consequence that contradicts the given relation $\angle p > \angle q$. Therefore the assumption that a is not greater than b is false.

Page 233, line 20: "Why?" If b' = b, then $\angle q' = \angle q$, by Principle 8.
But this contradicts the given relation $\angle q > \angle q'$.

Page 233: Theorem 31. If $\angle q$ is not greater than $\angle q'$, then either
$\angle q = \angle q'$ or $\angle q < \angle q'$. But each of these alternatives has a consequence
that contradicts the given relation $b > b'$. Therefore the assumption that
$\angle q$ is not greater than $\angle q'$ is false.

Page 234: Exercises.

1. The central angles corresponding to the two minor arcs are unequal.
   Apply Theorem 30.

2. Apply Theorem 31 and consider the minor arcs that correspond to the
   two angles of the triangles at the center of the circle.

3. In Fig. 10, $AB < BC$. Therefore $\frac{1}{2}AB < \frac{1}{2}BC$; that is, $MB < BN$. It follows
   from Theorem 28 that $\angle p > \angle q$. Consequently $\angle y > \angle x$ and $MO > NO$.

4. In Fig. 10, $MO > NO$. Therefore $\angle y > \angle x$, $\angle p > \angle q$, $BN > MB$, and $AB < BC$.

5. See Fig. A.

   $a + b > x$

   $c + d > x$

   $a + d > y$

   $b + c > y$

   $2a + 2b + 2c + 2d > 2x + 2y$

   $a + b + c + d > x + y$

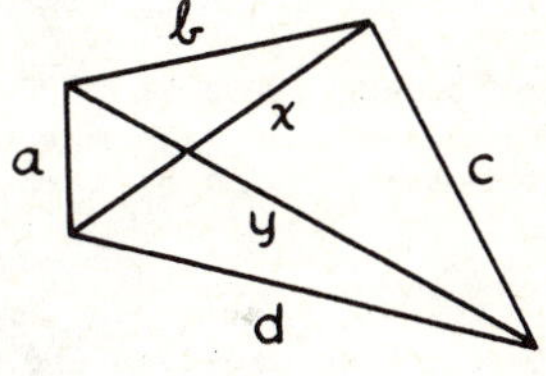

Fig. A

Pages 236-240: Exercises.

1. More exaggerated forms of Fig. 13a and Fig. 13h will show this.

2. In answering this question the pupil anticipates by his own efforts
   the series of diagrams shown in Fig. 13.

3. The directed angle $x_+ = \frac{1}{2}(\angle BOD_+ - \angle COA_+) = \frac{1}{2}(\angle BOD_+ + \angle AOC_-) =$
   $\frac{1}{2}(\widehat{BD}_+ + \widehat{AC}_-)$, where $\widehat{BD}_+$ is used to represent the measure of the directed central angle corresponding to the directed arc $\widehat{BD}_+$.

4. The directed angle $x_+ = \frac{1}{2}(\angle BOD_+) = \frac{1}{2}(\widehat{BD}_+)$. Since C coincides with A,

$\angle COA = \angle AOC = 0^\circ$. Consequently $\overparen{AC}$, intentionally printed without a subscript sign because $\overparen{AC}$ is zero, may be inserted in the parenthesis in order to preserve the form of the algebra. See page 236, lines 21-23.

8. The pupil must see that $360^\circ - \overparen{DA}_+$ can be replaced by $\overparen{AD}_+$.

10. The directed angle $x_+ = \tfrac{1}{2}(360^\circ) - y_+ = \tfrac{1}{2}(360^\circ - \overparen{CA}_+ + \overparen{BD}_+) = \tfrac{1}{2}(\overparen{AC}_+ + \overparen{BD}_+)$.

11. In circle i, $\tfrac{1}{2}(\overparen{BD}_+ + \overparen{AC}_+) = \tfrac{1}{2}(360^\circ)$.

   In circle j, $\angle x_+ = 180^\circ + \tfrac{1}{2}(\overparen{BD}_+ + \overparen{AC}_-)$, from Ex. 3. But $360^\circ + \overparen{AC}_- = \overparen{AC}_+$. Therefore $x_+ = \tfrac{1}{2}(\overparen{BD}_+ + \overparen{AC}_+)$.

   In circle k, $\overparen{AC}_+$ represents the measure of the central angle corresponding to the complete circumference, directed positively.

12. $\overline{PA}$ is positive, decreasing toward zero; $\overline{PB}$ is positive, approaching $\overline{AB}$; $\overline{PA} \times \overline{PB}$ is positive, decreasing toward zero.

13. Zero

14. $\overline{PA}$ is negative, decreasing algebraically toward $\overline{BA}$; $\overline{PB}$ is positive, decreasing toward zero; $\overline{PA} \times \overline{PB}$ is zero when P is at A, negative when P is between A and B; zero again when P is at B.

15. Zero

16. $\overline{PA}$ and $\overline{PB}$ are both negative and decreasing algebraically; $\overline{PA} \times \overline{PB}$ is positive and increasing.

17. Covered by answers to Exs. 12-16.

18. The perimeters of the rectangles in Fig. 14 are all equal; so the square has the largest area and the product rs is greatest when r = s.

   By the second suggested method, $\overline{AP} \times \overline{PB} = \left(\dfrac{\overline{AB}}{2} - \overline{PM}\right) \times \left(\dfrac{\overline{AB}}{2} + \overline{PM}\right) = \dfrac{(\overline{AB})^2}{4} - (\overline{PM})^2$. This has its greatest value, namely $\dfrac{(\overline{AB})^2}{4}$, when $\overline{PM} = 0$. Therefore the largest negative value attained by $\overline{PA} \times \overline{PB}$ is $-\dfrac{(\overline{AB})^2}{4}$. This occurs when P is at M, midway between A and B.

<u>Page 243, line 16</u>. The locus is a circular cylinder of radius 2 in.

<u>Page 244, line 1</u>. The locus is composed of four straight line segments, each equal in length to a side of the square, and four quadrants of a circle whose radius is equal to the radius of the rolling circle.

<u>Pages 244-246: Exercises</u>.

1. A circle, center at O, radius 5 inches.

2. Two parallel lines, each 4 inches from the given line. When the fixed line is perpendicular to the plane, the locus is a circle of radius 4 inches.

3. The four points in which the circle with center O and radius 5 inches intersects the two lines that are parallel to AB and 4 inches from AB. These four points are the corners of a rectangle, 8 inches by 6 inches.

4. The three points common to the circle and to the two lines that are parallel to AB and 4 inches from it. One of these lines is tangent to the circle. These three points are the vertices of a triangle of base 8 inches and altitude 8 inches.

5. A straight line midway between the two given lines.

6. Two lines parallel to the given lines, one on each side of the plane of the given lines and distant $\sqrt{5}$ inches from this plane.

7. Two lines, each parallel to the base of the triangle and at a distance equal to the altitude.

8. A line perpendicular to the chord and midway between one extremity of the chord and its perpendicular bisector.

9. Two lines, each making an angle of 30° with the given line.

10. A straight line perpendicular to the diameter (extended) through P. This straight line meets the diameter extended at a point D such that $OD \cdot OP = r^2$, where O is the center of the circle and r its radius. All that is expected of the pupil is that he shall plot

enough points of the locus to surmise that it is a straight line.
As P approaches O, the locus recedes from O; when P is at O, the
locus has vanished.

The proof, which is not expected of the
pupil, will be of interest to the teacher.
The similar right triangles OAD and OPA in
Fig. A tell us that $\frac{OD}{r} = \frac{r}{OP}$. Another pair
of similar triangles tells us that $\frac{OT}{r} = \frac{r}{OM}$.
Therefore OD x OP = OT x OM and $\frac{OD}{OT} = \frac{OM}{OP}$.
Since triangles ODT and OMP have angle MOP
in common and the sides including this angle
proportional, the two triangles are similar. Consequently angle ODT
is equal to angle OMP, which is 90°.

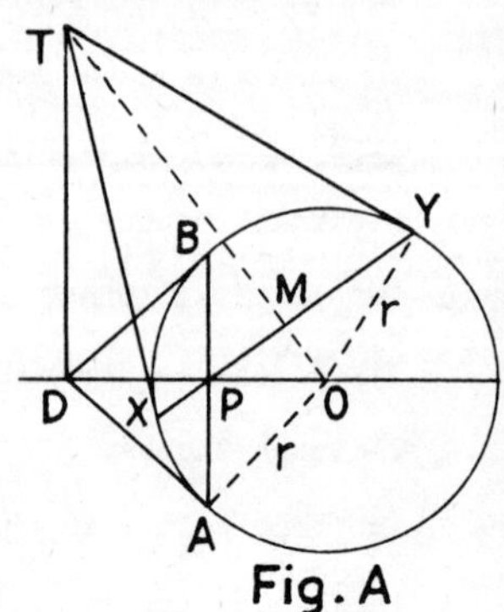

Fig. A

This locus can be thought of as the inverse of the circle on OP as
diameter, using the given circle as circle of inversion. See pages
263-264. So considered, it involves the converse of the theorem on
page 264, lines 17-27, namely Ex. 2 on page 265.

11. If one side of the square is s, the locus is a quadrant of a circle
of radius s, a quadrant of a circle of radius $s\sqrt{2}$, and another quadrant of a circle of radius s.

12. A circle concentric with the given circle and of radius 8. See Fig. B.

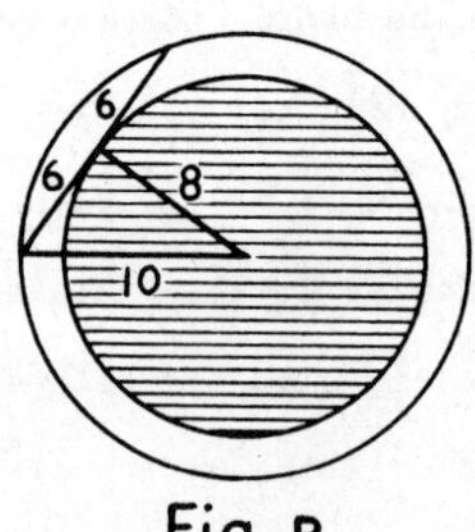

Fig. B

13. A parabola. Of course, the pupil is not expected to know anything
    about the curve he has plotted, not
    even its name. The teacher can
    tell him. See Fig. A.

14. Both branches of a hyperbola. Many
    students will draw only the right-hand branch.
    See Ex. 13 and Fig. B.

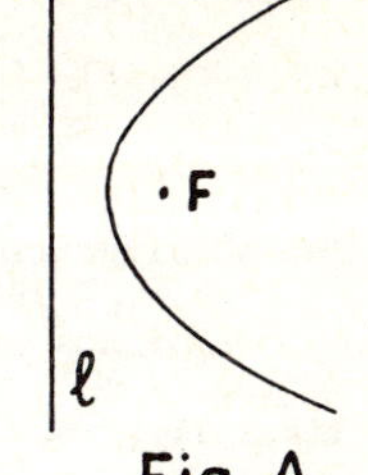

15. An ellipse, as shown in Fig. C.

16. Four straight line segments, each 1 inch long, and four quadrants of
    a circle of radius $\frac{3}{2}$ inches. In Fig. D, M is the mid-point of the
    hypotenuse of a right triangle; consequently M is at a distance $\frac{3}{2}$
    inches from the vertex of the right angle.

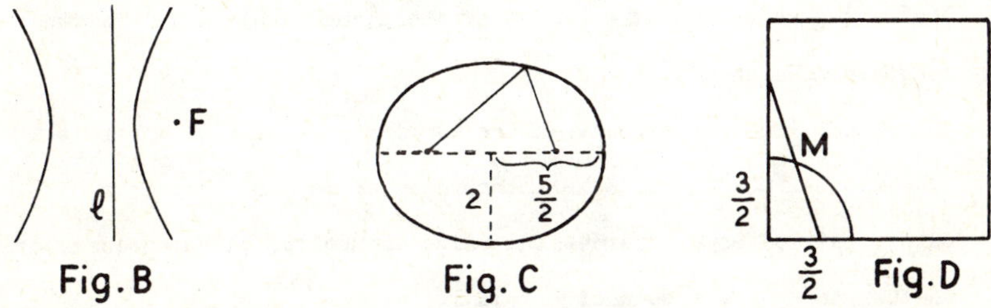

Page 249, line 22. First "Why:" by Corollary 14a. Second "Why:" by
Ex. 2 on page 113.

Page 250, lines 8-9. The locus is a plane perpendicular to the plane
of the given parallel lines, parallel to them and midway between them.

Page 251, line 19. The locus is a plane perpendicular to the line
segment joining the two given points and bisecting this line segment.

Page 252, lines 13-16. In each case the locus is a pair of planes
which bisect the angles formed by the given lines or planes.

Pages 254-261: Exercises.

1. The perpendicular bisector of the base.

2. A straight line midway between the two given parallels.

3. A straight line parallel to the given line and midway between it and
   the given point.

4. Two parallel lines, one on each side of BC, both equally distant
   from it.

5. Two straight lines through A, each making an angle of 45° with AB.
   Most students will think of AB as horizontal and will construe "upper"
   literally. Let them suppose that AB is vertical.

6. A straight line perpendicular to XY at P.

7. A circle having the given point as center and the given radius as
   its radius.

8. The perpendicular bisector of the line segment joining the two points.

9. A circle concentric with the given circle. Its radius will be
   $\sqrt{r^2 - \frac{l^2}{4}}$, where r is the radius of the given circle and l is the
   length of the chords.

10. A straight line midway between the two fixed parallels. This is
    true whether the exercise is interpreted as meaning that each circle
    cuts a pair of equal chords, the pairs themselves being unequal, or
    whether the pairs also must be equal.

11. Both the center of the circle and the point of intersection of the
    two tangents are equidistant from the points of contact. Hence they
    must lie on the perpendicular bisector of the chord of contact.

12. Since each mid-point is equidistant from the ends of the chord, the
    two mid-points must lie on the perpendicular bisector of the chord.

13. The center and the mid-points of the arcs are all equidistant from
    the ends of the chord; consequently they must lie on the perpendicu-
    lar bisector of the chord.

14. The center of each circle is equidistant from the ends of the common
    chord. Hence the two centers must lie on the perpendicular bisector
    of the common chord.

15. By Ex. 13, the center of the circle lies on the perpendicular bisec-
    tor of each chord. But the perpendicular bisector of one chord must
    be perpendicular to the other chord also, so the two perpendicular
    bisectors must coincide.

16. The diameter perpendicular to any chord of the system, by Ex. 15.

17. Draw any two chords. Their perpendicular bisectors will meet at the
    center.

18. The perpendicular bisectors of two sides AB and BC of any triangle
    ABC cannot be parallel; for if they were, the two sides would be
    parallel, or coincident. It follows that the two perpendicular bi-
    sectors must have a point in common. This point must be equidistant
    from A and B, and equidistant also from B and C. Since it is equi-
    distant from A and C, it must lie on the perpendicular bisector of
    the third side, AC, also.

19. The bisectors of two angles A and B of any triangle ABC cannot be
    parallel; for if they were, angles A and B would add up to 180°. It
    follows that the two bisectors must have a point in common. This
    point must be equidistant from AB and AC, and equidistant also from
    BA and BC. Since it is equidistant from AC and BC, it must lie on
    the bisector of the third angle C, also.

20. In contrast with the construction on page 181 of BASIC GEOMETRY, the
    emphasis in this exercise is now on the phrase "and only one." The
    proof follows immediately from Ex. 18 in this set of exercises.

21. Ordinarily the point of intersection of the first and second perpen-
    dicular bisectors will not coincide with the point of intersection
    of the second and third perpendicular bisectors.

22. A segment of the bisector of each angle of the triangle, each segment
    extending from vertex to incenter. See Ex. 19.

23. A pair of lines parallel to the given lines. If the distances from

the given lines l and m respectively are in the ratio p to q, one of

the new lines will be at the distance $\left(\dfrac{p}{p + q}\right)$ d from l and the other

will be at the distance $\left(\dfrac{p}{q - p}\right)$ d from l, where d is the distance

between l and m. The second part of this answer can be got by solv-

ing the equation $\dfrac{y}{y + d} = \dfrac{p}{q}$.

24. With P as center draw a circle that will cut the given circle in two

points A and B. Draw the perpendicular bisector of AB. Finally,

draw at P the perpendicular to this perpendicular bisector.

25. A circle concentric with the given circle and having a radius equal

to $\sqrt{r^2 + t^2}$.

26. An arc of the circle that has for its diameter the line segment join-

ing the center and the given external point. Every point of this

arc is inside the given circle.

27. The circle that has for its diameter the line segment joining the

center O of the concentric circles and the given external point P.

Points O and P do not belong to the locus.

28. A circle concentric with the given circle and having a radius of

10 feet.

29. Two equal circles, each having half the base for its diameter. The

mid-point of the base does not belong to the locus.

30. Same locus as in Ex. 29, except that now the end-points of the base

are also excluded from the locus.

31. A circle having for diameter the line segment joining the given point

and the center of the given circle. The given point does not belong

to the locus.

32. Same locus as in Ex. 31, except that now the given point is included

in the locus.

33. A circle with center at the intersection of the two fixed rods and

with radius equal to $\dfrac{1}{2}$, where l is the length of the moving rod.

The mid-point M of the moving rod is always the mid-point of the hypotenuse of a right triangle and so is always the same distance, $\frac{1}{2}$, from the ends of the moving rod and from the point of intersection of the two fixed rods.

34. Assume that the theorem is not true. It is still possible to pass a circle through three vertices of the given quadrilateral; the fourth vertex will be either inside or outside the circle. The two angles that are given as having the sum 180° correspond to two central angles that add up to 360°. But under the assumption that the fourth vertex is not on the circle, the given supplementary angles correspond to two central angles that add up to something more than or less than 360°: a contradiction. Therefore the fourth vertex must be on the circle.

35. A circle having AB as chord. Its center Q will be outside the given circle, on the perpendicular bisector of AB, and such that angle AQB is equal to 180° minus the central angle in the given circle corresponding to the minor arc AB. The proof depends on Locus Theorem 7 and Exs. 5 and 6, pages 147 and 148.

36. The phrase "segment of a circle" has not been defined previously; the description in parenthesis is sufficient to show its meaning.

If O is the center of the given circle, if P is the point of intersection of AC and BD, and if P' is the point of intersection of AD and BC, then angle APB is always 30° and angle AP'B is always 90°.

(a) The locus of P is an arc of a second circle having AB as chord and such that the minor arc AB of this second circle has a central angle of 60°. This means that the center Q of the second circle is "above" AB, on the perpendicular bisector of AB, and such that angle AQB is equal to 60°. Consequently Q is on the given circle, twice as far above O as O is above AB. The extent of the arc that constitutes the locus of P can be determined as follows.

One limiting position of CD makes C coincide with A. In this position $\angle$ BAD = 90° and BD extended meets the second circle at A', so that $\angle$ BAA' = 120° and A'Q is parallel to AB. The other limiting position of CD makes D coincide with B. In this position AC extended meets the second circle at B', so that $\angle$ ABB' = 120° and QB' is parallel to AB. The straight line A'QB' is tangent to the given circle and is a diameter of the second circle. Except for the end-points A' and B', all points of the second circle "above" this diameter constitute the locus of P. That is, the locus is a semicircle minus its end-points.

(b) The locus of P' is an arc of a third circle having AB as chord and such that arc AB of this third circle has a central angle of 90°. This means that the center R of this third circle is the mid-point of AB, and the locus of P' is the "upper" semicircle of this third circle, including the end-points A and B.

37. Each of the given triangles has its vertex V on one of two equal arcs of the sort shown in Fig. 11, page 253, in connection with Locus Theorem 7. For all positions of V on one of these arcs the desired locus is the complete circle that contains the other arc, except for that point on the major arc AB that is equidistant from A and B. This excepted point is approached on each side by the point of intersection P of the perpendiculars as V approaches first A and then B. But this excepted point cannot belong to the locus because V cannot coincide with either A or B.

When $\angle$ VAB = 90°, A is seen to belong to the locus. Similarly for B, when $\angle$ VBA = 90°. When V is between these two positions, $\angle$ APB is the supplement of the given angle. When V is outside these two positions, $\angle$ APB is equal to the given angle.

The entire locus, then, is made up of the two equal circles containing the arcs to which vertex V is always restricted, except for the point on each circle that is farthest from AB.

38. The line segment joining the mid-point of the base to the opposite
vertex. This line is defined on page 259 as a _median_ of the triangle.

39. The center will be at the point where the bisector of the 114° angle
intersects a parallel to one of the given lines that is 100 feet
distant from this given line.

40. For every position of CD angle C is unaltered in size; similarly for
angle D. Consequently angle DBC must be constant also.

41. Drop AD perpendicular to MR and continue it to A' so that A'D = AD.
The intersection of A'B and MR is the desired point.

42. Same as Ex. 41.

43. In Fig. 21 on page 259 of BASIC GEOMETRY, quadrilateral ABCB' is a
parallelogram in which $\angle$B' = $\angle$B, AB' = BC, and AB = B'C. Similarly
for quadrilaterals BCAC' and CABA'. Consequently triangle A'B'C' is
similar to triangle ABC and sides A'B', B'C', and C'A' are bisected
by C, A, and B respectively. So the altitudes of the given triangle
are the perpendicular bisectors of the sides of the new triangle,
and hence (by Ex. 18, page 255) meet in a point.

44. The suggestion given in the exercise is enough.

45. The point of intersection of each pair of tangents is on a bisector
of an angle of the triangle formed by joining the centers of the
three circles. Since it must be on all three bisectors, it must be
the point that is common to all three bisectors (by Ex. 19, page 256).

46. The locus consists of two lines, which
can be constructed as follows. Draw
a line parallel to AB and two units
away from AB. Draw two lines at a
distance of one unit from AC, one on
each side of AC. These two lines
will intersect the first line at P

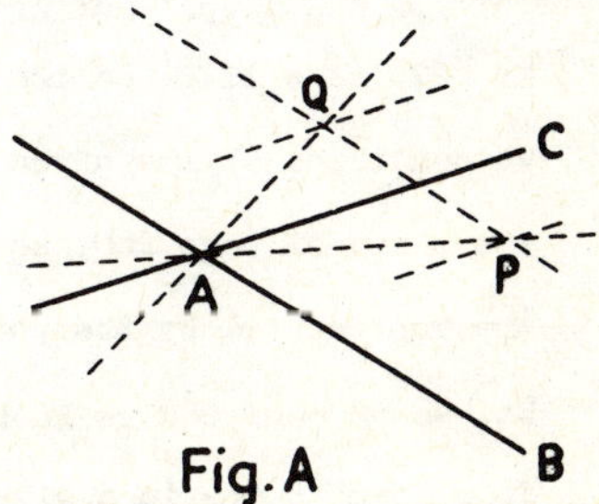

and Q, as in Fig. A. The lines AP and AQ, extended, constitute the
locus.

It is easy to prove that every point on AP is twice as far from AB
as from AC, and that every point on AQ is twice as far from AB as
from AC. In each case one needs only two pairs of similar right
triangles.

The difficulty in this exercise consists in proving the converse,
namely: if $\frac{PB}{PC} = \frac{2}{1}$ and if $\frac{P'B'}{P'C'} = \frac{2}{1}$, then P' must lie on AP (extended).
Angles BPC and B'P'C' are equal, since
each is the supplement of angle A.
Consequently triangles BPC and B'P'C'
are similar, and $\angle PBC = \angle P'B'C'$.
It follows that BC and B'C' are parallel, since each meets AB' at the same
angle. Therefore $\frac{AB}{AB'} = \left(\frac{BC}{B'C'}\right) = \frac{BP}{B'P'}$;
the right triangles ABP and AB'P' are similar; $\angle BAP = \angle B'AP'$;
and P' lies on AP.

The proof for Q and Q' follows the pattern of the preceding proof
for P and P' with only one change: angles BQC and B'Q'C' are now
equal to angle A instead of to its supplement.

47. $\frac{AR}{RB} = \left(\frac{AP}{PB}\right) = \frac{QA}{QB} = \frac{m}{n}$. See note following Ex. 26 on page 116.

48. It is evident from the preceding exercise that R and Q are two points
on the desired locus. Any other point P on the locus must be such
that $\frac{AP}{PB} = \frac{m}{n} = \frac{AR}{RB} = \frac{QA}{QB}$. This means that $\angle QPR$ must equal 90°, by
Ex. 27, page 117. So the locus of P is the circle on QR as diameter.

49. We can think of the given parallel lines as being perpendicular to
the plane of page 260, so that these lines - when viewed end-on -
are represented by the points A and B of Fig. 23 on that page. From
Ex. 48 on page 260 we know that the locus of points whose distances
from A and B are in a given ratio is the circle that has QR for

diameter. So in this Ex. 49 the locus must be a cylinder with axis through the midpoint of QR and perpendicular to the plane of page 260. That is, the locus is a cylinder with axis parallel to the given parallel lines.

50. The locus is a circle in the given plane with center D and radius 3. For every point in the plane that is 5 inches from P must be 3 inches from D.

51. The intersection is a circle with center at D, the foot of the perpendicular dropped to the plane from P, the center of the sphere. See Fig. 24, page 261. For if the radius of the sphere is r, every point of the intersection of plane and sphere will be at the same distance, $\sqrt{r^2 - (PD)^2}$, from D.

52. Every point of the intersection of two spheres with centers O and O' will lie in a plane perpendicular to OO'. See Ex. 14, page 143. So the intersection of the two spheres can be regarded as the intersection of this plane and either one of the spheres. By the preceding exercise this is a circle.

<u>Page 262: Exercises.</u>

1. The proof follows immediately from the definition of <u>power of a point with respect to a circle</u> on page 261. For every point P on the common chord, and on the common chord extended, the product PA · PB is the same for both circles.

2. This is a limiting case of the preceding exercise. If the circles are externally tangent at T, the power of any point P on the common internal tangent is $(PT)^2$ with respect to both circles.

3. If the circles are internally tangent at T, the power of any point P on the common external tangent is the same, namely $(PT)^2$, with respect to both circles.

<u>Page 263: Exercises.</u>

1. The foregoing proof can be applied without alteration to the case of two intersecting circles. It follows from Ex. 1 on page 262 that each of the two points of intersection of the two circles has the same power with respect to both circles. Both of these points must lie on PD, therefore, and PD must be the common chord (extended).

2. Both in the case of two circles that are externally tangent and of two circles that are internally tangent, it seems reasonable and helpful to define the radical axis as the common tangent of the two circles.

3. A plane perpendicular to the line of centers of the two spheres. The student is not expected to be able to prove this. Actually the proof follows the same pattern as the proof in the case of two circles, on pages 262-263. Given a point P having the same power with respect to two spheres with centers at O and O'; Fig. 27 on page 262 can be considered as representing the section made by the plane POO', except that points T and T' ordinarily will not be in this plane. But the relations $(PO)^2 = (OT)^2 + (TP)^2$ and $(PO')^2 = (O'T')^2 + (T'P)^2$ hold just as before.

<u>Page 264, line 13</u>. The inverse of the circle of inversion is the circle of inversion itself.

<u>Page 264, line 16</u>. The radius of a circle with center at O times the radius of the circle that is its inverse is equal to the square of the radius of the circle of inversion. The centers of all three circles are at the center of inversion, O.

<u>Page 264, line 26</u>: "Why?" Because triangles OQP and OP'Q' are similar by the Principle of Similarity, Case 1, and angle OP'Q' is given a right angle.

Pages 264-265: Exercises.

1. The foregoing proof applies in each case without alteration. When
   P' is on the circle of inversion it coincides with P, and the radius
   OP' of the circle of inversion is the diameter of the circle that is
   the inverse of the straight line through P'.

2. The inverse is a straight line perpendicular to the line of centers
   OO' of the two circles. In Fig. A, let OP be the diameter of the
   given circle that passes through O. If P'
   is the inverse of P, and Q' the inverse of
   a random point Q on the given circle, then
   $OP \cdot OP' = r^2 = OQ \cdot OQ'$ and $\dfrac{OP}{OQ} = \dfrac{OQ'}{OP'}$.
   Therefore triangles POQ and Q'OP' are simi-
   lar (Principle of Similarity, Case 1) and
   $\angle OQP = \angle OP'Q'$. But $\angle OQP = 90°$, being
   inscribed in a semicircle. Therefore Q'P' is perpendicular to OP'
   at P', when Q is any point of the given circle except O or P. So
   the locus of the inverses of all points of the given circle is the
   straight line through P' perpendicular to OP.

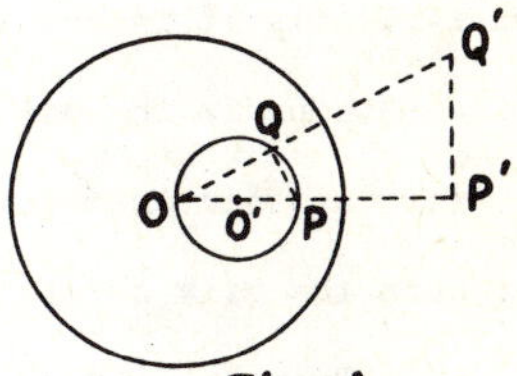

    Page 265, lines 10-11. A line segment equal in length to the diame-
ter of the circle.

    Page 265, lines 12-15. A circle equal to the circular edge of the
coin. Usually an ellipse; but when the two planes are perpendicular, the
projection is a line segment equal in length to the diameter of the coin.

    Page 265, lines 16-18. A circle. No.

    Pages 270-273: Exercises. As explained on page 269 the pupil is
not expected to find "the correct answer" to these exercises.

    Pages 274-276: Exercises.

1. No.

2. No. It might be an ellipsoid or other curved surface.

3. Keep it covered and chilled.

4. Cut a loaf into slices. Keep two slices dry and covered, but one
   warm and the other cold; keep two more slices moist and covered, but
   one warm and the other cold; keep two more dry and uncovered, but
   one warm and the other cold; keep two more moist and uncovered, but
   one warm and the other cold. Then observe which slice of each pair
   becomes moldy sooner. Test four more pairs with respect to moist
   and dry, and four more with respect to covered and uncovered.

5. Evidently it is not the air by itself that causes fermentation, but
   something in the air that is more commonly found in thickly settled
   regions than on mountain tops.

6. Heat the milk sufficiently to kill the ferment, or to kill most of
   it. Then chill the milk to discourage the growth of any of the ferment that remains alive. Also, keep air away from the milk.

7. The object is to drive out as much of the air as possible and then
   to kill the harmful bacteria that may be left inside the jars.

8. Because the pus-forming bacteria in the air were killed in passing
   through the carbolated gauze.

9. In the ice cream.   10. In the canned lobster.

<u>Pages 278-280: Exercises.</u>

1, 2, 3, 5, 6. In lines 17-19, on page 278, the student is reminded that
   he skipped the proofs of Principles 6, 7, 8, 11 and perhaps of Theorem 13 also. These proofs are given in the book on the pages men-
   tioned in these exercises.

4.   Given: Triangles ABC and A'B'C' (Fig. A) in which
   ∠A = ∠A', A'B' = k·AB, and A'C' = k·AC.

     To Prove: Triangle A'B'C'
   similar to triangle ABC.

     Proof: At B' draw B'C" so

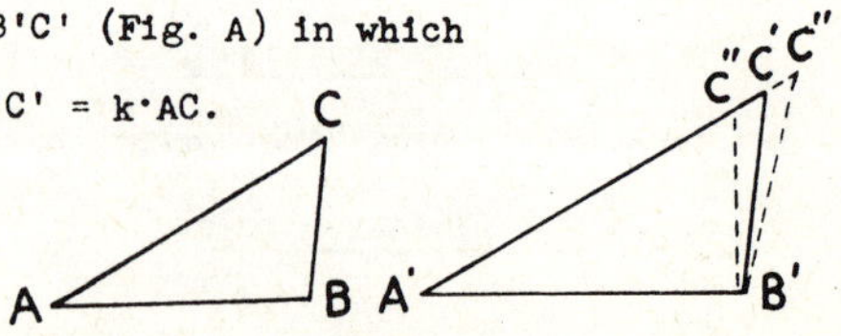

that $\angle A'B'C''$ equals $\angle B$. This line will meet $A'C'$ (extended beyond

the point C' if necessary) in the point C". By Case 2 of Similarity,

which for the moment is being taken as a fundamental postulate, tri-

angles ABC and A'B'C" are similar and $A'C'' = k \cdot AC$. But $A'C' = k \cdot AC$

(Given). Therefore $A'C'' = A'C'$ and C" must coincide with C'. It fol-

lows that $\angle A'B'C' = \angle B$ and triangles ABC and A'B'C' are similar.

    Case 3 can now be proved, just as on pages 79-80 of BASIC GEOMETRY.

7. Most of the answer is given on page 106 of BASIC GEOMETRY. The an-

swer there given and the form in which this Ex. 7 is worded both

imply that our chief interest here is in getting back to Principle 5.

Actually Principles 4 and 3 are also required in the proof of Theorem

13. Schematically the dependence of Theorem 13 upon these three

principles can be shown as follows.

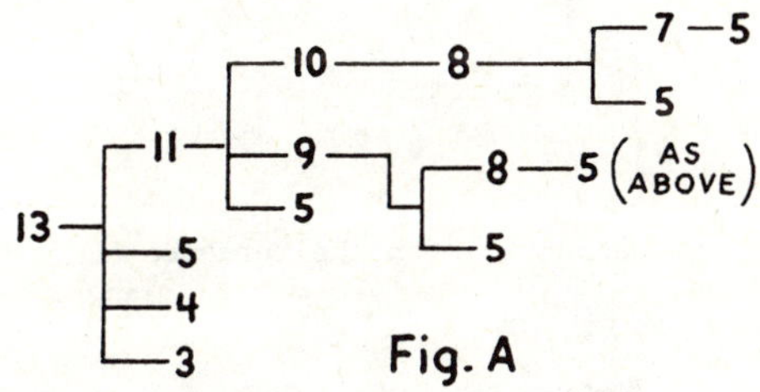

8. The dependence of Theorem 16 on Theorems 15, 14, and 13 and so back

to Principle 5 is shown in the following diagram.

9. The dependence of Theorem 20 on earlier theorems is shown in the

diagram on the next page (Fig. A).

10. On page 247 we have seen that if a proposition is true, its opposite
    converse is true also. So, instead of proving Theorem 21 directly

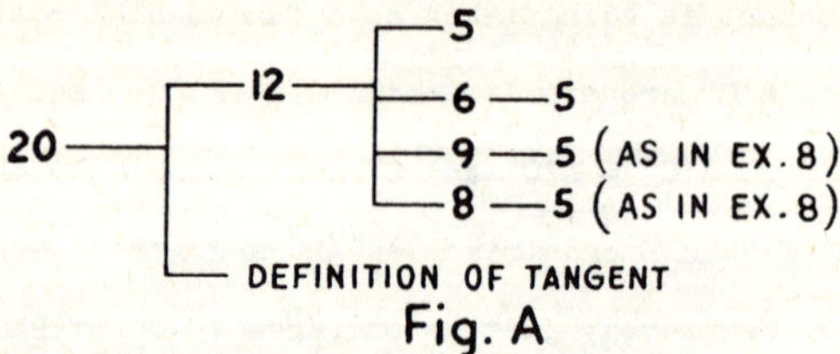

Fig. A

by showing that a given tangent is perpendicular to the radius, we
show instead that a line through T (page 139, Fig. 13) that is not
perpendicular to the radius OT cannot be tangent. This is as much
as is expected of the student.

11. The proof of Theorem 22 depends upon earlier theorems as indicated
    below.

Fig. B

12. The dependence of Theorem 23 upon the fundamental principles of this
    geometry is shown in Fig. C.

Fig. C

13. Any parallelogram having one side equal to b and the altitude upon
    this side equal to h can be divided into two triangles, each of side b

and altitude h, by drawing either one of the diagonals of the parallelogram. Consequently the area of the parallelogram is $2(\frac{1}{2}bh)$, or bh. Every rectangle is a special sort of parallelogram and so its area is equal to bh also. The area of any polygon can now be found just as described on pages 203-204 of BASIC GEOMETRY.

The contrast in procedure here is between the order rectangle - right triangle - any triangle - parallelogram - polygon and the order triangle - parallelogram (rectangle) - polygon.

14. In Fig. 9, page 32, of BASIC GEOMETRY, MO is the perpendicular bisector of AB and CO bisects angle ACB. $\triangle$ AMO = $\triangle$ BMO (by Case 3 of Similarity), and so $\angle$ MAO = $\angle$ MBO.

$\triangle$ ADO = $\triangle$ BEO (by the Pythagorean Theorem and Case 3 of Similarity), and so $\angle$ OAD = $\angle$ OBE.

Consequently $\angle$ MAO + $\angle$ OAD = $\angle$ MBO + $\angle$ OBE; $\angle$ BAC = $\angle$ ABC; and triangle ABC is isosceles.

15. Starting with triangle ABC in Fig. A in which CB $<$ CA, we wish to expose the fallacy in the foregoing "proof" that purports to show that CB = CA. Since the fault probably lies in the diagram shown in Fig. 9 on page 32 of BASIC GEOMETRY, we had better give careful consideration to the sort of triangle we draw.

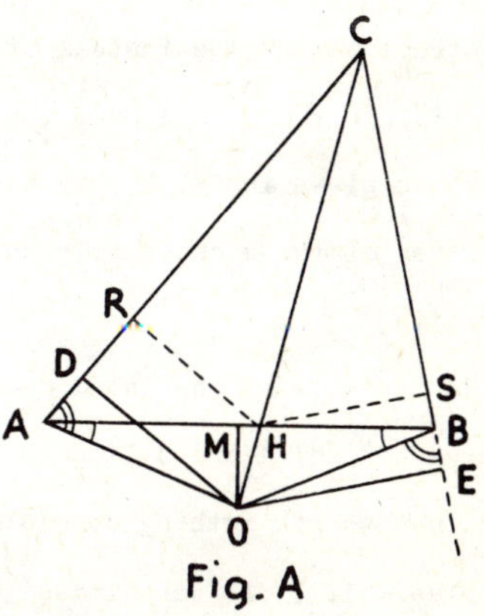

We know from Theorem 28 that $\angle$ CAB must be less than $\angle$ CBA. Consequently $\angle$ CAB must be acute and $\angle$ CBA might conceivably be acute, right, or obtuse. We dismiss the obtuse possibility at once, because if we should succeed in "proving" the theorem under this condition, $\angle$ CAB would have to be obtuse also, which is impossible. For the same reason we dismiss the possibility that $\angle CBA$ is a right angle.

Let us consider next whether the bisector of angle ACB intersects AB to the right or to the left of the mid-point M. If we draw perpendiculars HR and HS from H to CA and CB respectively, then $\triangle CHR = \triangle CHS$, CR = CS, and AR $>$ BS. Consequently, by the Pythagorean Theorem, AH $>$ HB and H is to the right of M. It follows that the intersection O of MO and CO must be outside the triangle, below AB, and not inside the triangle, as shown in Fig. 9 on page 32. This is the chief error in Fig. 9. If the student sees this, that is all that can fairly be expected of him.

Lastly, we must consider whether the perpendiculars OD and OE meet CA and CB respectively in two points D and E that are both between C and the corresponding vertex of the triangle; or both outside the triangle, on CA and CB extended; or one inside and the other outside. In the first case the purported proof appears still to hold if we subtract the angles instead of adding; namely $\angle OAD - \angle MAO = \angle OBE - \angle MBO$. In the second case the purported proof appears still to hold, just as given in Ex. 14, by adding the angles. Actually, however, neither of these cases is possible and the apparent proofs have no standing.

We turn now to the third case, shown in Fig. A, in which D lies between C and A and E lies on CB extended. This case is possible. But now we get nothing sensible either by adding or subtracting the angles. If $\angle CAB$ were indeed equal to $\angle CBA$, then $\angle OAD - \angle MAO$ would equal $180^\circ - (\angle OBE + \angle MBO)$. This would require that $\angle OAD$ should equal $180^\circ - \angle OBE$; namely, that the equal angles OAD and OBE should be supplementary, and consequently right angles. This would mean that triangles ADO and BEO would each contain two right angles, which is impossible.

Evidently we cannot prove triangle ABC isosceles so long as we
retain the initial condition that CA and CB are unequal.

Page 281: Exercises.

1. (a) Child saw cake.
   (b) Child did not eat cake.
   (c) Children see cakes.
   (d) Children will eat cakes.

2. (a) Man caught boy.
   (b) Man did not spank boy.
   (c) Men catch boys.
   (d) Men will spank boys.

3. (a) Circle was outside of triangle.
   (b) Circle was not inside of triangle.
   (c) Circles are outside of triangles.
   (d) Circles will be inside of triangles.

4. (a) Quotient was greater than divisor.
   (b) Quotient was not less than divisor.
   (c) Quotients are greater than divisors.
   (d) Quotients will be less than divisors.

Pages 282-283: Exercises.

1. Let us assume at the outset that $A < B$. When we shall have completed
   the proof under this assumption we shall need only to interchange A
   and B throughout to cover the case $B < A$.

   Given: $A < B$; $K < L$; $A < K < B$; $A < L < B$; and $K < X < L$.

   To prove: $A < X < B$.

   Proof: Since $A < K$ and $K < X$ (both given), we know
           that $A < X$ (Assumption 2).
           Since $X < L$ and $L < B$ (both given), we know
           that $X < B$ (Assumption 2).
           Therefore $A < X < B$ (Assumption 2).

2. Assumptions:
   (1.) A is older than B, or the same age as B, or younger than B.

(2.) If A is older than B and B is older than C, then A is older than C.

(3.) If A is the same age as B and B is the same age as C, then A is the same age as C.

<u>Theorems</u>:

(1.) If A is older than B and B is older than C and C is older than D, then A is older than D.

(2.) If A is the same age as B and B is older than C, then A is older than C.

3. <u>Assumptions</u>:

(1.) A is less than B, or equal to B, or greater than B.

(2.) If A is less than B and B is less than C, then A is less than C.

(3.) If A equals B and B equals C, then A equals C.

<u>Theorems</u>:

(1.) If A is less than B and B is less than C and C is less than D, then A is less than D.

(2.) If A equals B and B is less than C, then A is less than C.

4. <u>Assumptions</u>:

(1.) A precedes B, or coincides with B, or follows B.

(2.) If A precedes B and B precedes C, then A precedes C.

(3.) If A coincides with B and B coincides with C, then A coincides with C.

<u>Theorems</u>:

(1.) If A precedes B and B precedes C and C precedes D, then A precedes D.

(2.) If A coincides with B and B precedes C, then A precedes C.